Plate Tectonics

D. C. Heather

Edward Arnold

© D. C. Heather 1979

First published 1979
by Edward Arnold (Publishers) Ltd.,
41 Bedford Square, London WC1B 3DQ

Heather, D. C.
   Plate tectonics.
   1. Plate tectonics
   I. Title
   551.1'3   QE511.4
   ISBN 0–7131–0218–7

All Rights Reserved. No part of this publication may be reproduced, stored in a retrieval system, or transmitted, in any form or by any means, electronic, mechanical, photocopying, recording or otherwise, without the prior permission of Edward Arnold (Publishers) Limited.

Text set in 10/11 pt VIP Plantin, printed by photolithography, and bound in Great Britain at The Pitman Press, Bath

Acknowledgements

I wish to express my deep gratitude to Mrs Hazel Heather for typing my manuscript, to Dr Hugh Jenkyns of Oxford University for reading and commenting on the text and to my wife Catherine for her encouragement and assistance throughout its production.

Contents

1 **The interior of the Earth** 5
   The origin of the Earth 5
   The evidence for internal structure 6
   The internal structure of the Earth 10
   Effects of the Earth's internal structure 11

2 **Continental drift** 13
   Continental refits 13
   Further evidence for the former conjunction of continents 15
   The separation of the continents 24

3 **Plate tectonics** 26
   Sea-floor spreading 26
   Crustal plates 34

4 **Plates under tension (divergent plates)** 42
   Continental breakup 42
   The margins of new continents 45
   The expanding ocean floor 52

5 **Plates under compression (convergent plates)** 57
   Configurations of convergence 57
   Subductive convergence 58
   Collision convergence 63

6 **Continent building** 67
   The structure of age of continents 67
   The growth of continents 69

   Recommended Further Reading 77

   Index 78

Preface

The theory of 'plate tectonics' has, over the past ten years or so, radically altered the perspectives of the fields of both physical geography and geology. The theory postulates that the crust of the Earth is not continuous but rather a mosaic of separate plates. Moreover these plates are in motion relative to one another, constantly crashing together, tearing apart or sliding past each other. It is this tectonic activity which creates virtually all the major topographic features of the Earth's surface.

Since this theory has so recently been evolved it has not been comprehensively covered in the vast majority of physical geography books now in circulation in schools. Rather than providing a new basic text this book is intended to supplement those in current use. It is also intended to reveal to the reader the broad natural framework linking such features as continents and oceans; mountain chains and rift valleys; deep ocean trenches and volcanoes. It should also clarify the mechanisms of continental drift.

Existing texts on plate tectonics tend to fall into the categories of either introductions for the layman, middle school texts or university texts, none of which really suit the purposes of the sixth-year pupil. This book is intended to cover the material at just that level, incorporating many of the physical features in 'A' level syllabi.

Bound together by the theory, this material should be absorbed as a concept rather than learnt as a number of isolated topics. Diagrams have been kept simple so that they may be easily memorized. Where possible examples cited are located in the British Isles.

The main objective of this text is to bind the study of the major relief features of the Earth's surface into a cohesive pattern, and so to facilitate the examination candidate's task.

D. C. HEATHER

1

The interior of the Earth

The origin of the Earth

The origin of our own solar system has been deduced from observations of other such systems. One theory is that it originated from a flat spiral of dust and gases spinning in space. From this 'cloud', particles, attracted to each other by gravity, were drawn together to form a number of distinct bodies. These were the sun, the planets, the moons and asteroids of our solar system. The date of formation of the Earth, and also its moon, has been calculated as being about 4600 million years ago.

Studying the mode of formation of the Earth one can glean some indications as to its internal structure. At the time that this extraterrestrial debris came together to form the Earth the densest material gravitated to the centre whilst the less dense material accumulated around the outside. A comparable situation in a laboratory would be created by mixing sand, water and air in a sealed container, shaking them up and then allowing them to settle. The manner by which the components settle out, as shown in Fig. 1.1a, is called *density layering*. Due to its mode of formation the Earth possesses comparable density layers. As depicted in Fig. 1.1b these layers take the form of concentric shells.

The gaseous envelope making up the outer shell is called the *atmosphere*. The discontinuous shell of water inside that, the rivers and seas, is sometimes called the *hydrosphere*. The shell of rock inside that again is called the *lithosphere*. All these shells man

Fig. 1.1 Density layering: (a) sand, water and air in a bottle settling out in density layers; (b) the Earth settling out in density layers from a cloud of dust and gases spinning in space

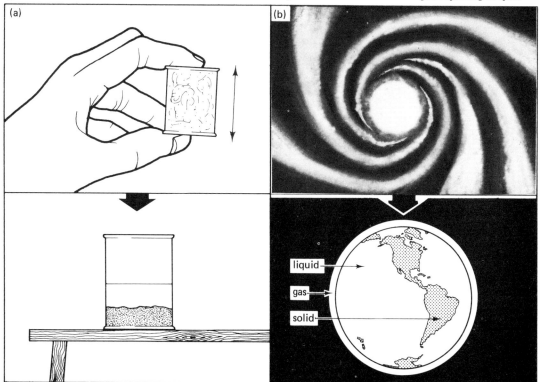

can observe easily. What he cannot do is see through the solid that lies beneath his feet.

Evidence for internal structure

Even the most advanced of drilling techniques cannot penetrate deeper than a few kilometres into the Earth's surface. If there are other, denser shells inside the one on which man lives, he must probe by indirect means. Direct evidence only extends as deep as can be drilled for samples. Indirect evidence comes from three sources: from *volcanoes*, from *meteorites* and from *seismic waves*. Although the evidence was not uncovered in this order chronologically, treating them in this sequence shows best the way in which these three sources of evidence combine together to provide a composite picture of the Earth's interior.

Volcanoes

Ever since man first ventured close enough to volcanic eruptions to see molten lava cool to form solid rock he might have conjectured that there might lie the very source of the rock on which he stood. It has long been speculated that volcanic products might give some clue as to the internal composition of this planet. A problem, however, stands in the way of such speculation. Different volcanoes in different parts of the world produce lavas with a variety of compositions.

Compare the two specimens of lava in plate 1.1. The specimen on the left is very pale in colour, being composed mainly of iron and aluminium silicate minerals. The specimen on the right however is very dark, being composed mainly of magnesium and iron silicate minerals. Which of these two types can then be claimed to represent the internal composition of the Earth?

One could postulate that these rock types evolved from a common ancestor, and perhaps developed along different lines while being transported from the Earth's interior to its surface. This view would be grossly simplistic in explaining all

Plate 1.1

variations in the compositions of *igneous* rocks (i.e. those which have solidified from the molten state). Some studies do indicate, however, that the longer molten material, or *magma*, takes to reach the surface, the more opportunity it has to alter its form and adjust to changing conditions of pressure and temperature. Therefore if samples of material known to have ascended rapidly from depth could be found they should not have changed markedly from their original state. These could then be said to represent material found deep below the Earth's surface.

A suitable eruption was recorded in Hawaii. The lava there contained lumps, or *xenoliths*, of a dark green, very dense rock. This proved on analysis to be a rock type known as *peridotite*. Additional evidence to support the theory of a peridotite-type interior to our planet comes from an extra-terrestrial source.

Meteorites

From time to time fragments of space debris become trapped in Earth's gravity field and come plummeting down as meteorites. Most of these meteorites burn up with the friction of passing through the atmosphere but some survive and crash to the Earth's surface. Meteorites which have been retrieved to date can be broadly divided into two categories: *stony meteorites* and *metallic meteorites*.

Until recently it was widely believed that meteorites originated from the breaking up of a hypothetical planet which orbited somewhere between Mars and Jupiter. This is the region now occupied by the asteroid belt. However, modern research indicates that asteroids are not a shattered planet but rather solid material that never gravitated together to form a planetary body in the first place.

The composition of the stony meteorites bears a marked resemblance to that of peridotite. One might then make the supposition that this is the type of material which went to make up the interior of our planet. This still leaves the metallic meteorites to be accounted for. These are usually made up of various *iron/nickel alloys*, and have considerably higher densities than their stony counterparts. The fate of the metallic meteorites at the time of the Earth's formation has then to be explained. For an explanation one may turn again to theories on density layering.

If the interior of the Earth is composed of these two materials it is reasonable to assume that the denser of the two would have gravitated to the centre. One could therefore assume the Earth to have a *core* of iron/nickel alloy, wrapped in a *mantle* of the less-dense peridotite-type rock. A rough preliminary model of the Earth's interior could therefore be constructed as in Fig. 1.2.

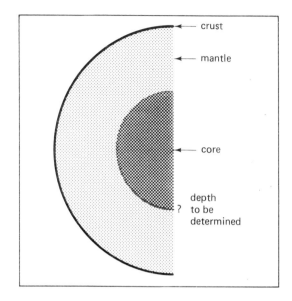

Fig. 1.2 A preliminary model of the Earth's interior

Making so many assumptions may seem a bit tenuous but this assumed model would fit in well with long-standing calculations of the Earth's overall mass. A great deal more evidence to support this model comes from another source; the study of *seismic waves*.

Seismic Waves

Seismic waves are vibrations in the Earth produced by any shock movements occurring within it. The commonest forms of movement are those which produce *earthquakes*. These occur when two sections of the Earth's surface layer, or *crust*, slide past each other. The sliding movement is not smooth, but jerky due to the effect of friction between the two sections. This effect can be observed by pressing one's hand down very hard on a table surface and sliding it along. The hand should advance by a series of sudden slipping movements, which jar and shake the table. In an earthquake this type of sudden movement causes vibrations in the Earth, which are known as seismic waves. The effects of these waves may vary from being barely detectable to being catastrophic, depending on the violence of

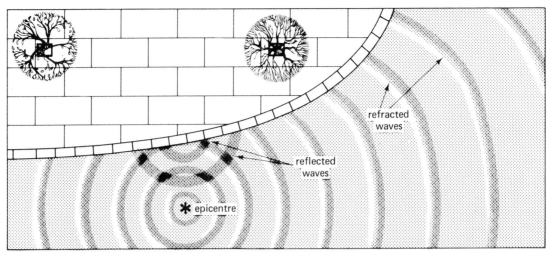

Fig. 1.3 Wave motion in a pool of water

the movement, and hence the magnitude of the earthquake.

An analogous situation can be observed where stones of various sizes are thrown into a pool of still water. The smaller stones cause small ripples which die out not far from their point of origin. The larger stones set up waves which may travel right to the edge of the pool. Fig. 1.3 shows how these waves may be reflected or refracted.

An earthquake is sometimes called a *seismic event*. The area of movement and the point the waves radiate out from is called the *epicentre*, or *focus*. If a seismic event is of sufficient magnitude the waves radiating out from the epicentre can pass right through the Earth. A study of the passage of these waves shows that, like the ripples in a pond, they can be reflected or refracted as they pass through different layers deep inside the planet.

The analysis of the routes followed by these waves is made by studying seismograph recordings. The seismograph is a very sensitive instrument designed to respond to the Earth's vibrations. It detects seismic waves and transfers information about them to a trace on graph paper. One recording in isolation has a comparatively limited value, but a composite picture, built up from the recordings of many seismographs, can be very valuable indeed (Fig. 1.4).

There are several different types of naturally occurring waves in the Earth, but the two principal types are known as '*P*' *waves* and '*S*' *waves*. The P wave is so named because it is the first, or *primary*, wave to be detected by the seismograph. P waves are by nature *pressure* waves, similar to sound waves. Pressure waves can travel through both solids and liquids, though they travel with much greater speed through solids. Their speed also increases in proportion to the density of the material through which they travel. The significance of this is that by calculating the speeds that P waves have travelled something can be learned of the density and state of the material through which they have travelled.

The S waves, or *secondary* waves, arrive at the seismograph after the primaries. They are by nature *shear* waves. These are up and down vibrations of a type that can only travel through solids. If an iron bar is held in the hand and then struck with a hammer this is the type of vibration that stings the hand. Since S waves can only travel through solids, should they meet a liquid layer inside the Earth they would be stopped short. Fig. 1.5 shows how this happens.

In Fig. 1.5 a cross-section of the Earth is shown which cuts right through the epicentre. The S waves can be traced by seismographs as they radiate out from the epicentre. At 103° from the epicentre, however, they suddenly disappear.

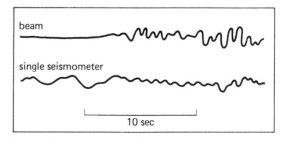

Fig. 1.4 A trace from a single seismograph compared to that from an array of 525 seismographs, beamed at the epicentre

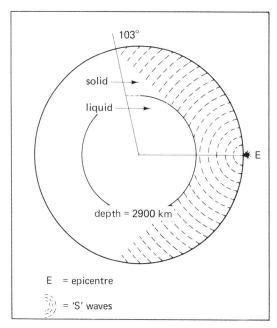

Fig. 1.5 How the depth of the liquid core is detected by the cutting-out of 'S' waves

Fig. 1.6 The shadow zone cast by the Earth's core in the case of an earthquake originating in Japan

From this it is deduced that at a depth of 2900 km the material inside the Earth changes from solid to liquid.

Since S waves are cut out by this liquid, to discover what lies deeper than 2900 km P waves must be further utilized. P waves can pass through the liquid and probe still deeper into the Earth. Although P waves can travel through the liquid they are *refracted* en route, which is to say their passage is distorted. The result of this is that from 103° to 142° from the epicentre, they are not detectable. In effect a seismic shadow is cast in this zone, as seen in Fig. 1.6. Beyond 142° however, the P waves can once again be detected, recorded and timed.

Recordings made by seismographs close to the epicentre have detected P waves being *reflected*, or bounced back, from some feature deep within the Earth (Fig. 1.7). By calculating the speed at which P waves travel the feature was estimated to occur at a depth of 5120 km. From the velocity of P waves below this depth the feature was determined to be the transition from liquid back to solid.

One other major feature has been detected by both P and S waves. From a depth of 75 km to 400 km below the surface both wave types travel more slowly. This is interpreted as being due to a semi-liquid layer occurring between these depths.

From the evidence gleaned from seismic waves then a composite picture of the Earth can be built up. This picture is outlined in the following section.

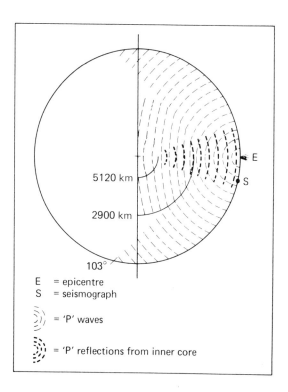

Fig. 1.7 Reflection of 'P' waves shows up the position of the inner core

9

The internal structure of the Earth

The core

At the centre of the Earth lies a *core*, the inside of which is solid and the outside of which is liquid. The velocity at which P waves travel through the core indicates that it has a density similar to that of iron/nickel meteorites. On this basis it is believed that the core of the Earth is composed of an iron/nickel alloy.

The mantle

Wrapped around the core lies a *mantle*, the inside of which is solid and the outside of which is semi-liquid. The speed at which P waves and S waves travel through the mantle indicates that it has a density similar to that of peridotite and also that of stony meteorites. On this basis it is believed that the mantle is composed of a dense rock similar in composition to peridotite. A. E. Ringwood of Cambridge University, proposing a hypothetical rock type to best suit all known information, has called this *pyrolite*.

The crust

On the outer surface of the mantle lies a *crust* of less dense rock. Living on the crust, man can examine the top of it by drilling into it for samples. With his very limited technology, however, this is still barely scratching the surface. Seismic waves again provide the main source of information. They mark clearly the bottom of the crust because they undergo a sudden increase in velocity, indicating that they have entered the denser rocks of the top of the mantle. This boundary is called the Mohorovicic discontinuity, after its discoverer Mohorovic. It is usually shortened to 'Moho'. The depth of the Moho varies considerably. This means that the depth of the crust varies accordingly. This variation is most marked between the crust underlying the continents and their margins, and the crust underlying the ocean floors. The crust beneath the ocean floors is thinner, denser and younger than that underlying the continents and their margins. For this reason the crust is divided into two types: oceanic crust and continental crust.

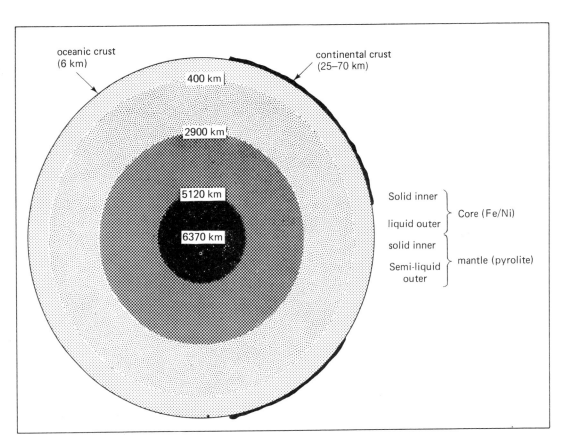

Fig. 1.8 The interior of the Earth

The *oceanic crust* is believed to be 8–9 km thick. Considering the radius of the Earth to be 6370 km it is comparatively very thin and therefore, as shown in later chapters, very susceptible to damage from stresses within the Earth. The crust's average density is 3 gm/cc, though seismic studies show that it is made up of four different layers with varying densities. Being composed mainly of *mag*nesium *si*licate minerals it is sometimes known as *sima*.

The *continental crust* varies in thickness from 25–70 km, but the average is about 35 km. The average density is 2.85 gm/cc. Being composed largely of *al*uminium *si*licate minerals it is sometimes known as *sial*.

The third point of contrast between continental and oceanic crust is that of age. The oldest known rocks retrieved from the ocean floor are 200 million years old. This is almost recent compared to the oldest known continental rocks, which have been dated at 3800 million years. These factors are indicators of the impermanence of oceanic crust in contrast to the relative stability of continental crust.

Using this more detailed information a useful model of the Earth's interior can be constructed as shown in Fig. 1.8.

The effects of the Earth's internal structure

The effects of this internal layering of the Earth are twofold. Firstly, motion between the layers may create the *Earth's magnetic field*. Secondly, the structure of the upper mantle and crust provides the setting for plate tectonics.

The Earth's magnetic field

The Earth's magnetic field is somewhat similar to that of a dipole magnet, as shown in Fig. 1.9. This is a somewhat simplistic view, however, since the Earth's field is a great deal more complex. Temperatures inside the Earth also are far too high for a simple magnet to survive. An alternative to producing a field by means of a magnet is to produce one from an electric current.

A German physicist, Walter Elsasser, suggested that the Earth might act like a dynamo to produce electric currents in the following fashion. Chemical reactions in the core might act like a battery to produce a weak electric charge. This initial charge could then be acted on by eddies in the liquid outer core, which might function as a dynamo to amplify it. Some type of feedback would be necessary to

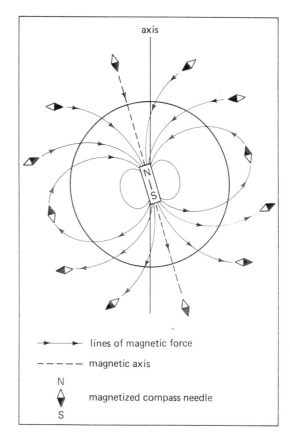

Fig. 1.9 The Earth's magnetic field. It resembles one that would be produced by a dipole magnet in the core. Compass needles would align themselves along the flux lines.

prevent the charge, and hence the field, from dying out so the dynamo would have to be of a self-exciting type. Fig. 1.10 shows a resemblance in form and motion between suggested eddies in the outer core and a self-exciting dynamo. They might both perhaps resemble rotating discs.

There are many drawbacks to this anology but it does have some points in its favour. Firstly, it provides a mechanism which could produce an electro-magnetic field. Secondly, it would explain why the Earth's field is aligned roughly parallel to its axis. Thirdly, it provides a system which is sufficiently flexible to explain small variations with time of the position of the field. Although it remains roughly aligned with the axis it does shift. At London for instance the compass north position has swung through 35° in 240 years. Fourthly, it is known that the Earth's magnetic field undergoes *reversals* of polarity from time to time (i.e. North becomes South and vice-versa). Modification of the dynamo model could account for this. As will be

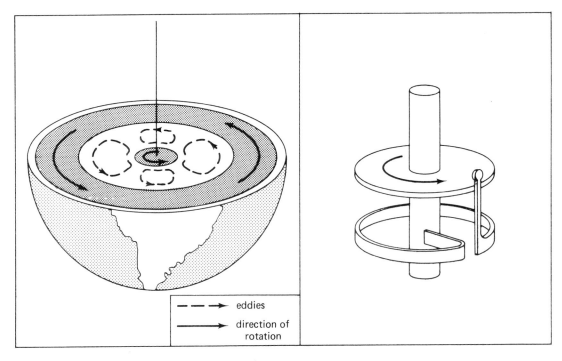

Fig. 1.10 Relative movement between the mantle and the inner core creates eddies with a movement similar to that of the self-exciting dynamo shown on the right

seen later these reversals have great significance in the study of plate tectonics.

The setting for plate tectonics

Both S and P waves, as they cross the Moho, undergo a marked increase in velocity. However, at a depth of about 70 km the velocity of both waves suddenly drops off. In the S waves the slowing down is much more marked. This is interpreted as meaning that the rocks at that depth are partially molten. Because of its effect on seismic waves this region is known as the *low velocity zone*. An alternative name given to it is the *rheosphere* since the semi-liquid rock is *rheid*, or capable of flowage.

Any flowage in the rheosphere is bound to have a considerable effect on the material overlying it. It would be at least distorted, or might even be moved about. The evidence showing that this does in fact happen was first examined by scientists studying *continental drift*.

2

Continental drift

Continental refits

Early refits

Looking at a Mercator projection of the world in an atlas it is easy to sense that one is looking at a great natural jig-saw puzzle. Francis Bacon speculated as early as 1620 on the manner in which opposing shore-lines could be fitted together. By 1858 Antonio Snider had published a book in Paris which contained the map shown in Fig. 2.1. This shows his re-assembly of the continents for about 300 million years ago. In 1910 an American, F. B. Taylor, published his ideas on continental movement. He described land masses flowing outwards from the poles. Some of his argumentation was easily refuted, and because of this his ideas as a whole were generally rejected by the academic world.

The first writer to publish a really strong case for continental movement was not even a geologist but a German meteorologist by the name of Alfred Wegener. His book *The Genesis of Continents and Oceans*, published in 1915, included maps like that shown in Fig. 2.2. This particular example shows how Wegener explained evidence for glaciation in lands that are now tropical. He re-assembled them about the South Pole, again, at a time about 300 million years before the present.

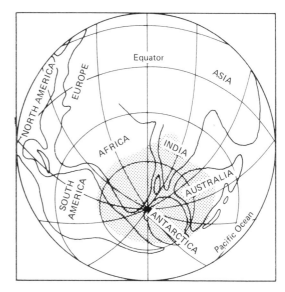

Fig. 2.2 Map showing the distribution of glaciation 300 million years ago in Gondwanaland. After Alfred Wegener's 1915 map

The depth of his argument was such that some geologists did not lightly dismiss his ideas. Although some of these views, especially in regard to the forces moving the continents, were subsequently proved wrong, the impact of his work was sufficient to motivate further study of continental drift by other researchers.

Modern refits

Refits have become more sophisticated since those early days. The topographic map of the ocean floor shown in the centrefold shows clearly how refits

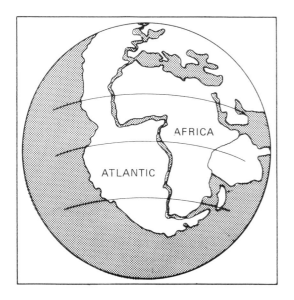

Fig. 2.1 Map published by A. Snider in 1855 showing his reassembly of the continents for a period 300 million years ago

13

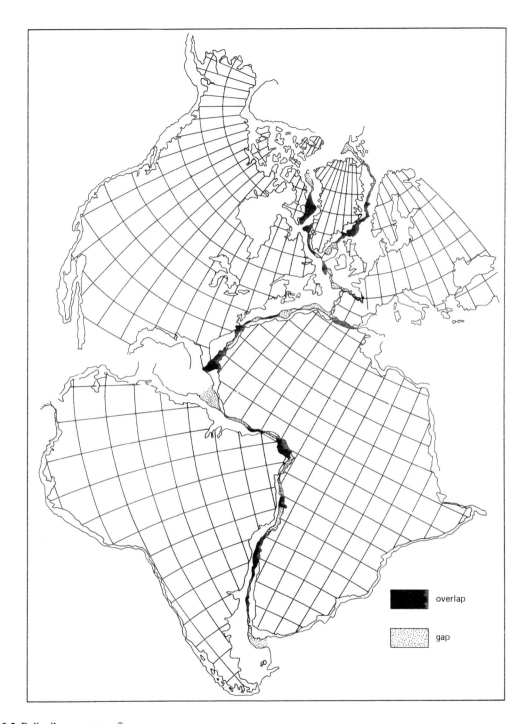

Fig. 2.3 Bullard's computer refit

should have been made using the edges of continental shelves rather than the shorelines of land masses. Using the continental shelf edges, which are the true continental margins, a team from Cambridge University, led by Sir Edward Bullard, produced, with the aid of a computer, the refit shown in Fig. 2.3. One can see that, although there are slight areas of mismatch, the refit is very close. The areas of overlap which occur can nearly all be explained as features which developed after the split-up.

Further evidence for the conjunction of continents

If the continents were at one stage joined together it is reasonable to assume that matching edges should have features in common which pre-date the separation. These features might relate to structure, to lithology (rock type), to rock magnetism, to former climates, or to fossils.

Structural evidence

The very hearts of the continents are composed of *basements* or *shields*. These, as shown in Chapter 6, are in part built up from the roots of ancient mountain chains, fused together. These *orogenic belts* are long and narrow and fairly clearly differentiated. They can also be reasonably accurately dated. On continents which have been torn apart, therefore, these belts should be traceable from one torn edge to the other. Fig. 2.4 shows the matching up of these belts between South America and Africa. The best area of match is between northeast Brazil and West Africa. This, however, could be due to the fact that these are the areas which have been studied in the most detail.

Similar evidence can be seen much closer to home. The chain of mountains which runs down through Scandinavia and Scotland can be geologically linked to the chain running from Newfoundland down through the Appalachians in the eastern United States.

These are only some of the orogenic belts whose courses in ancient times can be traced from continent to continent as shown on the refit map in Fig. 2.5. This is a reconstruction of the 'supercontinent' of *Pangaea*. (Named by Wegener, *pangaea* being the Greek for 'all earth'.) The continuity of the belts on the reconstruction is a convincing testimonial as to its accuracy.

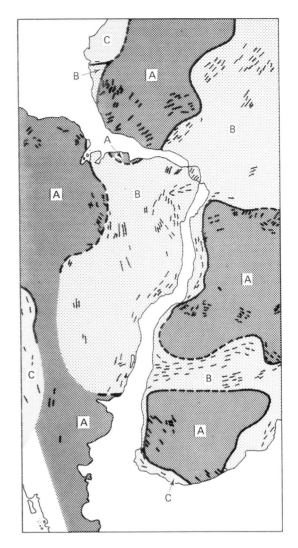

Fig. 2.4 Africa/South America refit showing continuity of structure. Areas labelled B are 550–100 million years old. A denotes areas older than B. C denotes areas younger than B. Short lines show structural trends in the metamorphic basement.

Lithological evidence

Lithology, or rock type, can provide evidence about continental drift. Where continents were once joined one would expect to find matching rock types on opposing sides of the 'join', except in the case of rocks which formed after the rift.

F. Ahmad of the Indian Geological Survey produced the map shown in Fig. 2.6. It indicates a link-up between the Permian rocks (230–270 million years old) between India and Australia. The lines show the thickness of the rock in thousands of feet. There is a clear correspondence in thickness

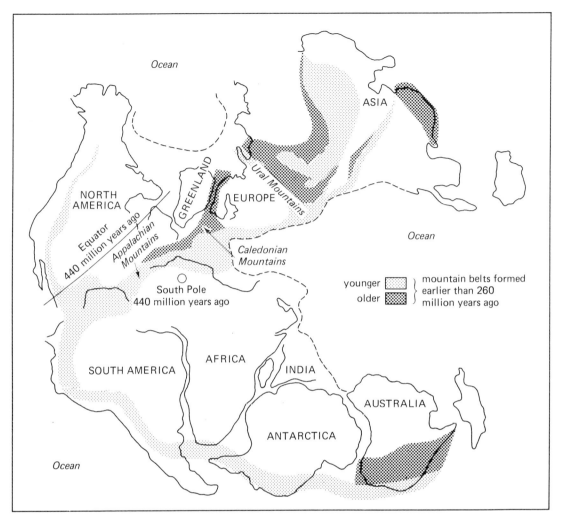

Fig. 2.5 The ancient continent of Pangaea is reconstructed by fitting together the major continental land masses. The mountain belts shown are the products of collisions between older land masses that came together to form Pangaea.

between the two countries. Not only this, but several very distinct bands of rock can be matched up. The finest example of this occurs where four beds of limestone composed almost entirely of the shells of a particular brachiopod are seen at Umaria: identical shell types abound in corresponding limestones on the west coast of Australia.

This is, however, only one isolated example. The case is further strengthened if, instead of matching just one rock band, or the lithology, one matches up the sequence of rock types, or the *stratigraphy*. Fig. 2.7 shows the comparative stratigraphy of the southern continents. The names on the side are those given to various geological divisions of time, with the younger ones placed at the top. As can be seen there is a high degree of correlation of rock types between the continents. This matching of rock types can, of course, only be made for the times when these continents were joined together. After splitting up, each continent developed its own particular stratigraphy.

This type of matching can be made, though not to the same extent, in the northern continents. Sedimentary and igneous rocks from Girvan and Ballantrae in Scotland, for instance, have been matched with similar rock-types in Newfoundland.

Palaeomagnetic evidence

The prefix *palaeo-* means ancient. In the context of palaeomagnetism it refers to the magnetic field acquired by a rock at its time of formation. This

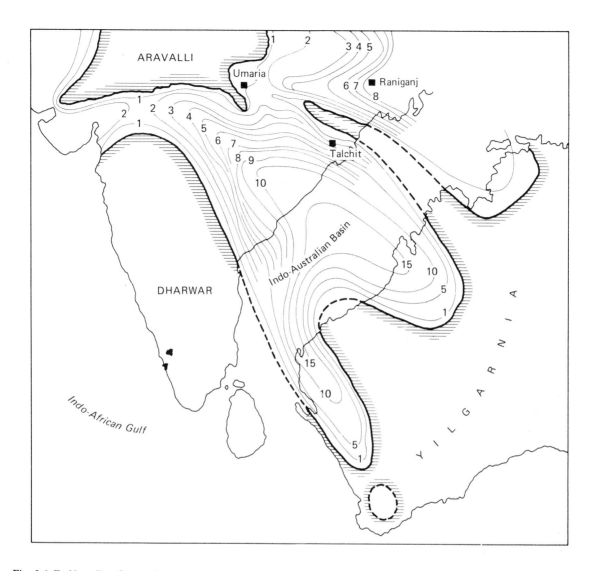

Fig. 2.6 F. Ahmad's refit map for India and Australia

field may be acquired by several means. In the case of igneous rocks as they cool past their Curie temperature (i.e. the temperature above which they cannot hold their magnetism) minerals in the rock may become magnetized in the direction of the Earth's magnetic field: in the case of sedimentary rocks any magnetic particles, as they are deposited, may align themselves parallel to the lines of the Earth's magnetic field. In either case if we know the age of the rock we can use this *rock magnetism* to determine the position of the Earth's magnetic poles at the time of the formation of that rock. To determine more exactly the pole positions one must take a collection of samples from several layers in a formation and from several different localities. This accommodates local minor variations and allows a statistical result to be achieved.

Deriving the positions of the North Pole in different continents at different times produces an unusual result. The North and South Poles appear to have moved considerably throughout geological time. We know that minor variations occur in the position of the Earth's field, but nothing on so large a scale as indicated by the geological record. This large scale movement is known as *polar wandering*.

If the positions of the North Pole are plotted at different intervals of time, taking measurements from one particular continent, a track can be plotted following the movement of the poles. This is called a *polar wandering curve*. Each continent, however, has its own individual polar wandering curve, as shown in Fig. 2.8. This leaves one with

17

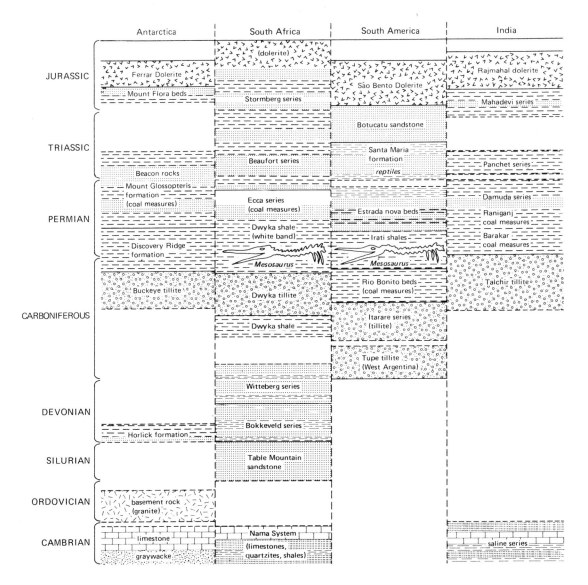

Fig. 2.7 Comparative stratigraphy of the southern continents

two alternative possibilities: either each continent has had its own North Pole, or else the continents have moved relative to one another. By plotting the various poles on a continental refit map (Fig. 2.9) one can see that they all fall within a relatively small area close to the South Pole. It seems logical therefore to choose the latter of the two possibilities, which is to say that the continents have moved, and that the magnetic poles have not shifted appreciably from their present-day alignment to the Earth's axis of rotation.

Palaeoclimatic evidence

The Earth's climate can be broadly divided into zones, from polar through to tropical. There seems no reason to suppose that *palaeoclimates*, or ancient climates, should not have been similarly restricted. If this were the case one might reasonably expect to find evidence reflected in rock types. Certain rock types only form under appropriate climatic conditions, and as a result of this their distributions are restricted.

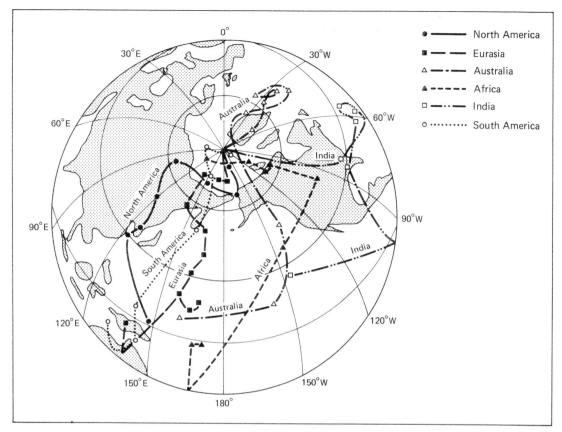

Fig. 2.8 Comparison of polar wandering curves of North America, Europe, South America, Africa, India and Australia

In Fig. 2.7 the rock sequences found in the southern continents are shown. At a similar level in each succession can be found *tillites*. Tillites are glacially deposited sedimentary rocks. They contain an unsorted mixture of particles, from clay to boulders. The larger fragments may bear grooves, or *striae*, where the moving ice has ground them against each other or over the bedrock. The bedrock too may be similarly striated. The striations on the bedrock can be used as indicators of the directions of ice movement. Wegener's theories on the glaciation of the southern continents were based on the distribution of tillites (Fig. 2.2).

A study in detail of areas of Africa and South America provides evidence that they were once linked. Fig. 2.10 presents information based on work by Martin, a South African geologist. It charts out areas of tillites and their depths (B). It also denotes areas which, at the same geological level, display widespread evidence of glacial erosion (A). It indicates also the direction of ice movement. If one takes these two areas as being separated at the time of glaciation one is presented with two problems. Firstly what became of the material stripped off broad areas of South Africa? Secondly, where lay the source of the great volumes of till deposited in South America? If one accepts the premise that the two areas were adjoined the problems cancel one another out. The material could have been ice-transported directly from A to B.

Limestones may be taken as indicators of warmer climates. *Coral*, one of the major sources of *calcium carbonate* (the 'lime' mineral in limestones), has its growth restricted to warm waters, and at present, growth is usually contained within 30° of the equator. Coralline rocks, however, are found in Lower Palaeozoic rocks within the Arctic Circle. In more recent rocks this zone of corals moves progressively south till it assumes its present position, spanning the tropics. The zone of coral growth therefore has either moved southward or, alternatively, the coralline rocks have been formed in the tropics and then carried northward on drifting

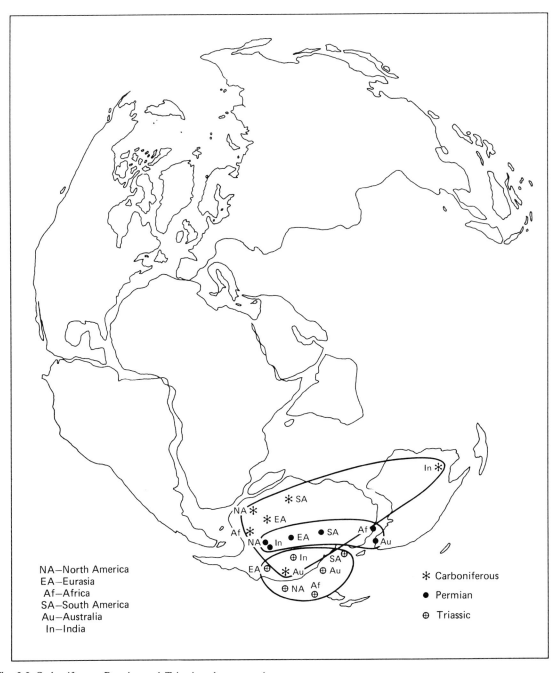

Fig. 2.9 Carboniferous, Permian and Triassic palaeomagnetic pole positions calculated for a Pangaea reconstruction

continents. Since corals could not have survived in the Arctic Circle anyway, due to lack of light, the latter theory is the only feasible alternative.

Another factor to be considered is that the precipitation of calcium carbonate, either organic or inorganic, does require warmth if it is to occur and form rocks on a significant scale. It would appear, then, that the formation of limestones occurs within the Earth's warmer regions and the rocks, once formed, may be transported to cooler areas.

Three rock types typify arid climates. These are

dune-bedded sandstones, laterites and evaporites.
Dune-bedded sandstones are the product of wind-blown sand and most of the world's present great expanses of sand occur in hot deserts, lying 10° to 40° from the equator. Dune-bedded sandstones can, however, occur outside these regions, and are best used as climatic indicators in conjunction with the other two types.

Laterites, or red beds, were once thought to be produced in moist tropical regions but study of the minerals in the rock now indicates that they, too, form in hot, arid regions.

The third of these types, *evaporites*, e.g. gypsum ($CaSO_4 \cdot 2H_2O$), rock salt ($NaCl$), sylvite (KCl), are formed under conditions where an area of salt-water is being evaporated more rapidly than it can be replenished. These conditions are best met in hot, arid regions.

These three rock types combine to give a good pattern of the extent of hot arid regions of the world at any particular time. Used in conjunction with glacial evidence to locate the poles, one could use this evidence to re-locate the continents at a particular period in geological time. In 1924 Wegener and the great German climatologist Koppen published a refit map based on this type of evidence. They also used the presence of coal as an indicator of moist conditions. The map in Fig. 2.11 is based on their 1924 map showing a refit with climatic zones for 280 million years ago. It is remarkably close to an up-to-date refit shown in Fig. 2.9. Palaeoclimatology, then, has a very important part to play alongside other techniques in establishing continental movement.

Palaeontological evidence

Another key for unlocking the mysteries of continental drift is found in the field of *palaeontology* which is the study of ancient species of life through the medium of fossil evidence. If, for example, one refers again to Fig. 2.7 one can see on the boundary between the Carboniferous and the Permian, (about 270 million years ago), the occurrence in shale bands of *Mesosaurus* fossils. *Mesosaurus* was a small aquatic reptile about five hundred millimetres long. Fig. 2.12 shows a restoration of *Mesosaurus* and indicates the regions where its fossil remains have been found. If this reptile could have swum well enough to cross the Atlantic it seems unlikely that its distribution would have been confined to this restricted zone. It might be concluded that the domain of *Mesosaurus* was limited by its swimming ability, and that this domain was

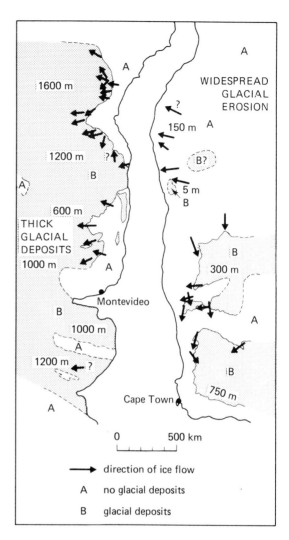

Fig. 2.10 Distribution and thickness in metres of Permo-Carboniferous (300 million years ago) glacial deposits in South America and Africa

rifted down the middle and the two halves were carried far apart.

An after effect of this type of event would be a *divergence of species*. This is to say that the descendants of a species split into two groups would continue to evolve but each group would evolve independently and therefore divergently from the other. The best example of this effect is seen in Africa, Australia and South America where, after breaking apart, each developed its own distinctive fauna.

The reverse of this effect, *convergence of species*, would be expected to occur where two land masses came into conjunction. In this situation one would

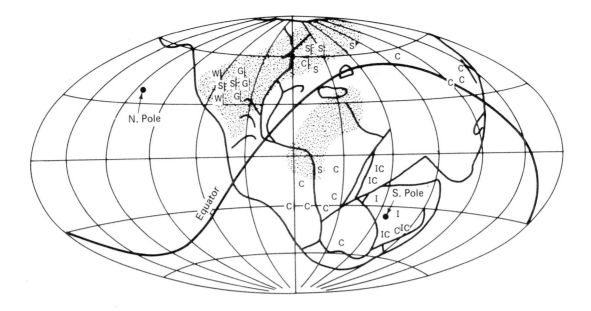

Fig. 2.11 Koppen and Wegener's Permian palaeoclimatic map (modified). The Pole and equator were estimated from the distribution of wind-blown sands (W and dotted pattern): rock salt (S) and gypsum (G); coal (C); glacial deposits (I)

expect the faunas of the two adjoining areas to circulate freely between them, and so develop a common fauna from their respective stocks. An effect to be noted here is that the weaker species, previously protected from competition by their isolation, may find themselves being squeezed out by fitter species from the other land mass. A case in point is seen in the union of North and South America. About 22 million years ago a junction, or *land bridge*, was established between the two regions. Prior to this time some 29 families of mammals lived in South America and some 27 different families lived in North America. After the union the two regions came to have 22 families in common. (See Fig. 2.13.)

This effect of convergence and divergence of species can be linked in many instances through geological time with the convergence and divergence of land masses. It should be noted that a land bridge, allowing the passage of terrestrial species, is a water barrier, preventing the passage of aquatic species. (See Fig. 2.14.) Thus the establishment of a land bridge causing convergence among land species will at the same time initiate

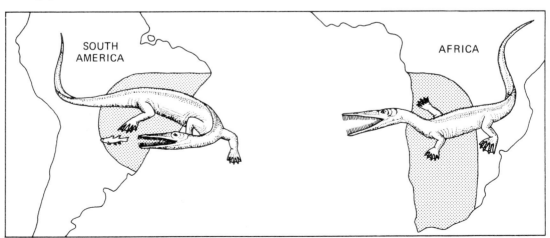

Fig. 2.12 Distribution of fossils of *Mesosaurus*

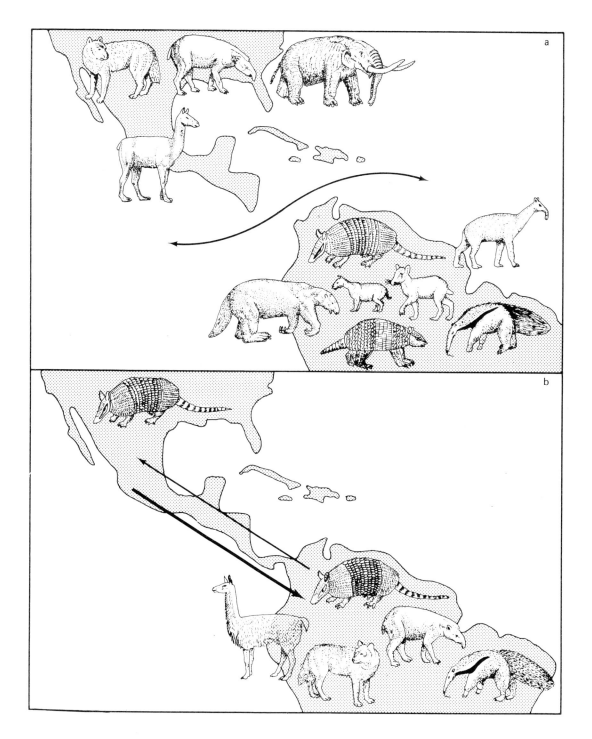

Fig. 2.13a Divergence of mammals in the New World occurred when North and South America were unconnected. Each developed its own distinctive fauna

Fig. 2.13b Convergence of mammals occurred when a 'land bridge' was established. The armadillo spread North. Many South American species were unable to compete with the northern species and became extinct

divergence in the divided marine population.

To sum up then, we can see how by studying fossil populations and noting the various evolutionary trends, the pattern of land movements through geological time can be established.

The separation of the continents

Early theories

Theories that continents had once been joined together were quite common in the nineteenth and early twentieth centuries. Several had been invoked by Victorian palaeontologists to explain the faunal links reviewed on page 21. These theories maintained that the continents remained static but that there were physical 'land bridges' in regions where, today, oceans exist. There was no question of land masses having moved. It was postulated that, like Atlantis, these 'bridges' had foundered and sunk without trace.

Two geophysical arguments counter these theories, both of which are based on the fact that the crust of the Earth is less dense than its mantle.

Firstly, it is extremely unusual for one material to sink into another, denser material. Secondly, if sections of the crust had sunk into the denser mantle their lower mass should have a lower gravitational attraction. Gravitational measurements, however, indicate that the reverse is the case; that the material of the sea floor is, in fact, *denser* than that of the continents.

Continental drift

Since theories of vertical displacement must be rejected on these grounds, theories involving horizontal displacement must be invoked. The body of evidence reviewed on page 15 indicates that horizontal movement does occur. One theory put forward suggested that the crust was stretched horizontally and so thinned to form the ocean floor. This, however, would not be in accordance with the known structure of the ocean floor. The alternative is to suggest, as Wegener did, that the continents were bodily transported laterally (i.e. that they drifted) to their present positions. If this is accepted it is necessary to explain the origin of

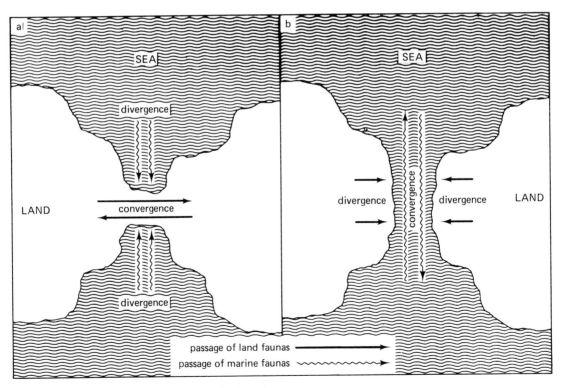

Fig. 2.14 Situation A causes convergence among land species and divergence among marine species. In situation B the process is reversed

the ocean-floor crust in the intervening space between the dislocated land masses. Magnetic, topographic and drilling surveys of the ocean floor since the early 1960s have provided the evidence to construct a hypothesis meeting this requirement.

3

Plate tectonics

Sea-floor spreading

The mid-ocean ridges

The centrefold map shows clearly how the continents stand out in relief above the ocean floor. Seismic surveys indicate that the crust underlying the oceans has an average thickness of only 8–9 km, in contrast with thicknesses of 25–70 km for continental crust. The ocean crust then is relatively thin. The map shows it to have several distinctive surface features. It is, however, relatively undistorted. Most of the ocean floor is covered by a thin veneer of sediments, and these, too, are known to be virtually undisturbed.

The most prominent structures seen on the sea floor are the *mid-ocean ridges*. These ridges form a long, continuous chain standing up to 3 km above the flanking plains of the sea floor. They have, in certain sections, median rift valleys and they follow a characteristic staggered path, being offset by a series of *transform faults*. The mid-Atlantic ridge is an outstanding example of such a ridge. It runs right down the length of the Atlantic Ocean, centrally balanced between Europe and North America; between Africa and South America. In the previous chapter it was shown how these lands were once joined together. The entire Atlantic floor then, including the ridge, must have been formed since they split apart.

One factor which points towards the general youth of ocean floors is that no sediments older than 160 million years have so far been retrieved from any part of them. In 1960 Professor Hess of Princeton University proposed a theory to explain this and other phenomena. He suggested that the mid-ocean ridges were the lines of generation for new crust. This is to say that as the continents split apart magma welled up along the line of the rift to cool and form new crust; that as they continued to separate they took with them, welded on, a trailing edge of fresh oceanic crust; and that the rift in their wake was constantly replenished by fresh upwellings of magma from the mantle below. In this manner the ocean floors grew outward from the central rifts, with the ridges being the surface expression of the upwelling magma. Two of the phenomena lending support to this theory are the heat flow and seismicity related to mid-ocean ridges.

Geothermal heat constantly seeps out through the Earth's crust from its interior. On the ocean floor this can be measured by means of a device called a thermistor probe. Readings from the ocean floor are generally similar to those from continents. Over the mid-ocean ridges, however, the readings may be several times the norm (Fig. 3.1). Hess interpreted this as being due to the heat produced by the injection of mantle material.

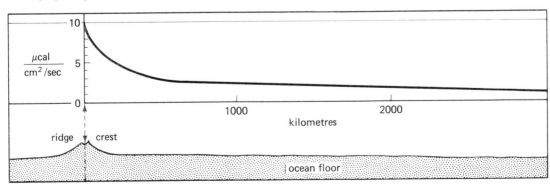

Fig. 3.1 Typical pattern of geothermal heat flow across a mid-ocean ridge

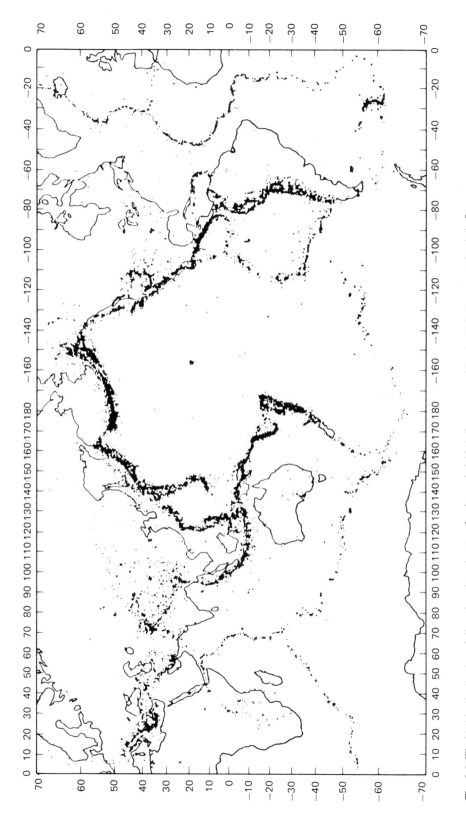

Fig. 3.2 Worldwide distribution of all earthquake epicentres for the period 1961–7, as reported by the U.S. Coast and Geodetic Survey

Seismic studies also indicate that the mid-ocean ridges are centres of activity. A comparison of the location of earthquakes, on the map in Fig. 3.2, with the relief of the ocean-floor shows a significant coincidence. The course of the mid-ocean ridge is clearly defined as an important seismic zone, and seismic activity is one function of crustal movement.

Although Hess put forward his theories on seafloor spreading in 1960 much of the confirming evidence for it was not amassed until the latter half of that decade. It had been discovered by the mid-sixties that the Earth's magnetic field frequently reverses its polarity (i.e. magnetic north becomes magnetic south and vice versa). This was deduced from worldwide studies of basalt lava flows. (Basalt is particularly efficient at acquiring a palaeomagnetic imprint.) A few years previous to this, in 1963, F. J. Vine and D. H. Matthews of Cambridge University had proposed a modifier for Hess's theory. They suggested that the uppermost part of the ocean's crust was basaltic, and that at the time of its formation it would have become magnetized in whichever direction the Earth's magnetic field happened to be orientated. This meant that as the ocean floor grew away from the ridge it would have acquired a fossil magnetism in the fashion shown in Fig 3.3. If this hypothesis were correct one would expect to find, on either flank of the ridge, matched bands of rock with either 'normal' (present) or 'reversed' polarity.

Marine magnetic surveys had, in fact, been undertaken from as early as 1950 and magnetic lineations had been recorded off the west coast of North America. No interpretation of them, however, had been undertaken at that time. Confirmation of the Vine-Matthews hypothesis came in 1966 when the results were published of an aeromagnetic survey of the *Reykjanes Ridge*, in the North Atlantic, south of Iceland (Fig. 3.4). This showed clearly the matched magnetic banding that Vine and Matthews had predicted. Later, similar surveys showed that this pattern of magnetic banding could be detected across most of the world's mid-ocean ridge system. These magnetic bands could also be dated, which was enormously significant, because it meant that rates of sea-floor spreading could be calculated and that the growth of the oceans could be retraced. Spreading rates vary considerably. At Iceland each side grows about 1 cm per year. In parts of the East Pacific Rise it may be up to 9 cm per year. In some areas, like the Carlsberg ridge in the Indian Ocean, magnetic banding cannot be differentiated at all. This might be because its rate of spreading is very slow, and produces such a tight banding that it is beyond the sensitivity of the equipment used in the surveys.

The dating of volcanic activity on the sea floor adds weight to the theory of sea-floor spreading. Iceland, to the north of the Reykjanes ridge, is the only point where a mid-ocean ridge protrudes above sea level. Fig. 3.4 shows how, in Iceland, the

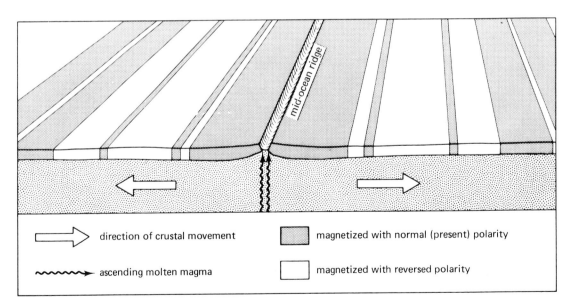

Fig. 3.3 Diagrammatic representation of the geomagnetization of the ocean floor

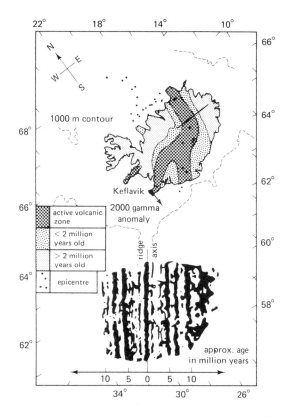

Fig. 3.4 The magnetic banding pattern on the Reykjanes Ridge and belts of volcanicity in Iceland. (Note youngest belt at centre.)

work of a purpose-built drilling ship, *Glomar Challenger*. With its special ability to drill sampling holes directly into the deep ocean floor *Glomar Challenger* rapidly produced firm evidence regarding the pattern of sedimentation on the sea-floor. It showed that the depth of sediment was related to distance from the ridge; that the age of the oldest sediments, those directly overlying the crust, was similarly related; and that the age of these oldest sediments matched the ages deduced from magnetic banding patterns.

If one accepts theories on sea-floor spreading then one is also obliged to accept one or other of two alternatives. Either: (a) the Earth has expanded to accommodate the increased area of crust; or (b) as fast as crust is being produced at one site it is being consumed at another. The first of these theories can be rejected on the grounds that, if the oceans have always expanded at their present rate, the volume of the Earth would have had to increase five-fold over the past 200 million years. There is most recent volcanic activity occurs in a band down the centre of the island and how, moving away to the East and West, the volcanics become progressively older. On a broader scale the dating of the volcanic islands in the North Atlantic (Fig. 3.5) shows that the farther from the ridge the islands are the older they are. This effect is easily explained in terms of sea-floor spreading. (Except for those volcanoes which have been produced later than the crust they overlie by hot spots in the mantle.)

Any sedimentation taking place on the sea-floor is bound to reflect evidence of spreading. On the newly formed oceanic crust on the flanks of the ridge one should find relatively little sediment, whilst moving farther from the ridge one should find progressively thicker sediments on older crust. Any section of the oceanic crust should be only slightly older than the sediments directly overlying it. In 1968 a venture was started with the object of testing this hypothesis. 'The Deep Sea Drilling Project' was a special programme based around the

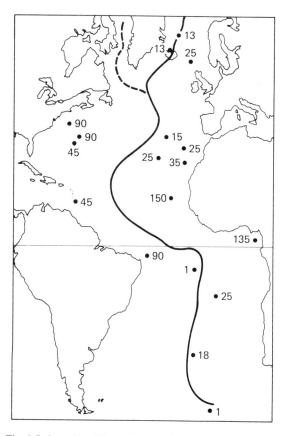

Fig. 3.5 Ages, in millions of years, of volcanoes in the Atlantic Ocean

Fig. 3.6 The island arc system of the West Pacific Ocean

strong geophysical evidence that this has not happened. This means that the second of these two alternatives is the more likely. The site of generation of new crust, the mid-ocean ridge, has been examined in this section. Possible sites for consumption of the old crust are examined in the next.

The deep-sea trenches

The distribution of earthquakes around the world is shown in Fig. 3.2. It can be seen that most of the activity is confined to distinct, linear zones. Some of these zones coincide with mid-ocean ridges. The most intense seismic activity, however, occurs elsewhere. A comparison with the centrefold map shows that the belts of high activity roughly underlie the *deep ocean trenches*. These are long, narrow troughs in the ocean bed, most of which occur in the Pacific Ocean. The deepest of these, the Mariana Trench (11 022 m deep) could engulf Mount Everest (8848 m high).

These deep-sea trenches are located in two distinctly different situations. The first of these is found where a trench fringes a continent. The classic example of this is seen in the Peru/Chile Trench which borders the west coast of South America. On the centrefold map it can be seen that the trench, the coastline and the Andean Cordillera all follow roughly parallel courses. Most of the earthquake foci in this belt underlie, not the trench, but the cordillera, which is also notable for a very high incidence of active volcanoes. In the second of these situations the trenches occur on the deep ocean floor, usually presenting a charactistic arcuate form (Fig. 3.6), and usually flanked by a *volcanic island arc*. Here, again, most of the seismic activity takes place, not under the trench, but under the island arc.

As an integral part of his theory of sea-floor spreading, Hess proposed that these trenches were the zones where crust was being thrust back down into the mantle, re-melted and thus re-absorbed.

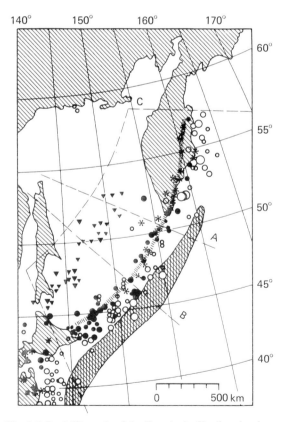
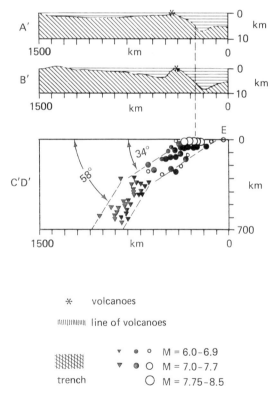

Fig. 3.7 Benioff's study of the Kamchatka-Kurile volcanic arc. Circles, dots and triangle represent respectively shallow, intermediate and deep earthquake foci; sizes vary according to magnitude as shown. A' and B' are exaggerated vertical sections across lines A and B. C'D' is a composite seismic cross-section of the arc from CC to DD

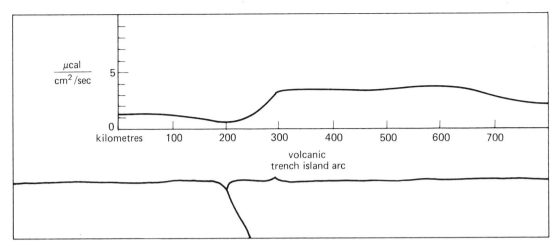

Fig. 3.8 Typical pattern of geothermal heat flow across an ocean trench and volcanic island arc

His thinking on this must have been influenced, not only by the global distribution of earthquakes, but by the detailed studies made of the seismic belts by Hugo Benioff in the 1950s. Benioff examined the depths of occurrence of earthquake foci. His results for the Kuril Trench are illustrated in Fig. 3.7. He discovered that earthquakes close to the trenches tended to be shallow whilst those further towards the island arcs, or the cordillera, tended to occur at progressively greater depths. He calculated that the average inclination of these zones was about 45°. These zones of descending foci have since been named the *Benioff Zones*. Their position in relation to trenches and island arcs can be seen on Fig. 3.6, where contours mark the depths of the Benioff Zones in hundreds of kilometres. Hess proposed that the Benioff Zones marked the line of disturbance caused by the passage of ocean crust as it was subducted, that is to say as it was thrust down into, and re-absorbed by, the mantle.

Evidence from other geophysical sources support theories suggesting the deep-ocean trenches as subduction zones. Just as geothermal heat flow is higher than the average on the mid-ocean ridges (Fig. 3.1), over the deep-ocean trenches it is lower (Fig. 3.8). This could be caused by the cold crust descending into, and thus cooling, the mantle. Measurements of the Earth's gravity field show that it, too, is slightly below the norm over the Benioff Zones. This indicates that the material below the surface is less dense than its surroundings. Ocean crust descending into denser mantle material could explain this effect.

Hess's hypothesis of sea-floor spreading provides an excellent framework for the study of geology

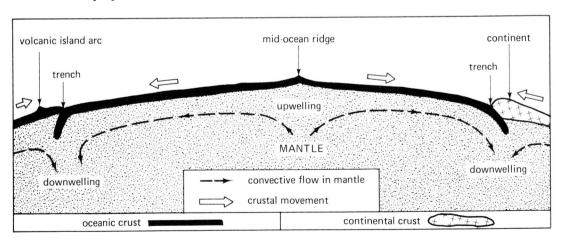

Fig. 3.9 Diagrammatic representation of Hess's model of sea-floor spreading

The Red Sea, created by Africa (right) tearing away from Eurasia (left)

and geophysics. Its contents are summarized in Fig. 3.9. New crust is generated by upwelling magma at mid-ocean ridges. This crust moves away from the ridges, carrying the continents with it. At the deep-ocean trenches it is re-absorbed into the mantle, plunging down under either the sea-floor or else a continent. However, the hypothesis has certain shortcomings. Hess explained his system as being driven by convection cells in the mantle, with ridges forming over rising flows and trenches over descending flows. This mechanism cannot explain the complex global movements of ridges and trenches shown in Fig. 3.10 which illustrates the occurrence of ridges and trenches around the equator. For various reasons Africa is believed to have held its present position for a considerable geological period. This means that the mid-Atlantic and mid-Indian Ocean ridges must be moving away from each other. They could not therefore hold static positions above rising limbs of convection cells. The theory also does not explain how large tracts of thin ocean crust have remained virtually undisturbed despite undergoing massive movements. The sea-floor spreading hypothesis therefore needs further modification if it is to explain these and other phenomena.

Crustal Plates

Plate structure

On page 8 the structure of the crust was examined through the medium of seismic studies. Fig. 3.11 shows a typical seismic profile through a section of oceanic crust and the structure of the crust as

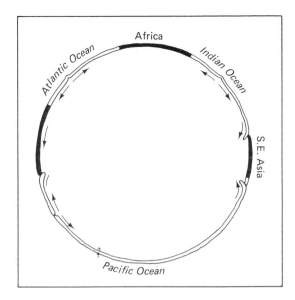

Fig. 3.10 Distribution of ridges and trenches around the equator. Continental crust is represented in black

interpreted from it. The Moho marks the base of the crust where the velocity of seismic waves increases as they enter the denser material of the mantle. At a depth of about 70–80 km the wave velocity decreases. The Low Velocity Zone is interpreted as being due to a layer of the mantle being in a partially melted state. The crust and the upper, solid part of the mantle are together known as the *lithosphere*. The partially melted material of the L.V.Z. is known as the asthenosphere.

If the Earth's surface moves about independently of its interior, then the boundary between the

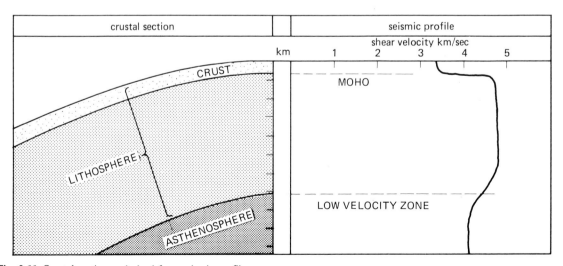

Fig. 3.11 Crustal section as derived from seismic profile

lithosphere and asthenosphere is the most likely site for de-coupling to take place. This is to say that the crust is rigidly attached to the upper part of the mantle and that these two together move about the surface of the globe, riding on top of the semi-solid asthenosphere. Instead of a thin skin of crust bounding the Earth, as envisaged in Hess's and earlier theories, it is enclosed in the strong case of the lithosphere; a case which is broken into a number of separate, rigid *plates*. This explains why thousands of square kilometres of ocean crust have remained undeformed and why the continents can still be neatly interlocked like the pieces of a jig-saw puzzle. The crust merely forms the topmost layer of these plates and as they move about it is carried along on top of them in a pick-a-back fashion.

Continental crust is much thicker than oceanic crust—about 35 km as compared to 8.9 km. Also it is less dense, averaging 2.85 gm/cc as compared to 3 gm/cc. The effect of these differences is that areas of plate capped by continental crust are both *more rigid* and more *buoyant* than those capped by oceanic crust. Most plates are capped by both types of crust. These are classed as *continental plates*. Those which are capped entirely, or almost entirely, by oceanic crust are classed as *oceanic plates*. The continental parts of plates, because of their greater rigidity and buoyancy, resist subduction back down into the mantle. For this reason the crust of the continents is about 1000 million years old but can be as much as 4000 million years old.

The thinner, weaker and denser oceanic crust, on the other hand is relatively easily dragged down and re-absorbed into the mantle. The most ancient oceanic crust known is no older than 200 million years.

Plate boundaries

Since plates are relatively thick and therefore strong, they are internally fairly rigid. As they move about relative to one another, however, their margins come under conditions of considerable stress. In response to this stress, rock may either bend (fold) or it may break (fault). *Folding* usually occurs under conditions of sustained stress, while *faulting* usually occurs where the rock is subjected to violent stress. The dislocation which occurs during faulting produces shock waves. The pattern of seismic waves produced depends on the nature of the dislocation, which in turn is the product of the type of stress producing it. Fig. 3.12 shows how rock will fault in response to different types of stress. The arrows show the directions of displacement, which are different for each type of faulting. This causes each type of faulting to produce its own identifiable pattern of seismic waves. Analyses of seismic patterns therefore can be used to determine the type of motion occurring at plate boundaries.

Hess's theories on sea-floor spreading were substantially correct, except that not just crust, but lithospheric plates, are being created and destroyed

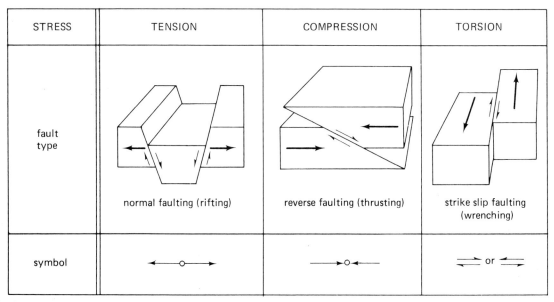

Fig. 3.12 Fault dislocations produced by differing stress fields

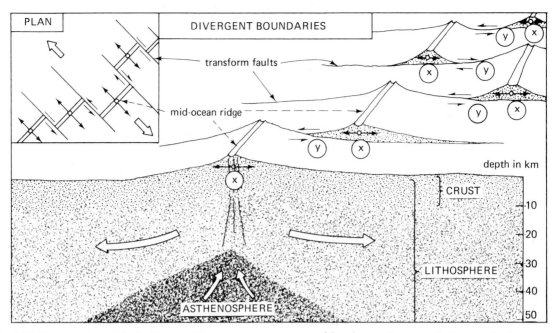

Fig. 3.13 Plate tectonic model of mid-ocean ridge

at the ridges and trenches. In the light of this it is necessary to re-examine the structure of plate boundaries. Here three types of relative motion are possible. Plates may be pulling apart (tension), pushing together (compression) or sliding past one another (torsion).

According to the sea-floor spreading hypothesis one would expect to find conditions of tension along the line of the mid-ocean ridge. Seismic studies show that the asthenosphere comes closer to the surface along the ridge than elsewhere. This would be in accordance with ideas about the ridge being a line of upwelling of magma. It would also explain the high geothermal heat flow over the ridge (Fig. 3.1) and why earthquakes along the ridge have shallow foci. As there is a rift in the crest of much of the ridge one would expect to find the shock pattern produced by normal faulting. This is in fact the case. Also, a second type of shock pattern occurs; that produced by strike-slip faulting. This occurs along the line of the faults which transversely offset the ridge. The reason for this is seen in Fig. 3.13. Although the two plates are diverging from each other, at the points marked (y) they will be, due to the offsets by the faults, sliding past each other. They transform one type of relative motion into another and because of this they are called *transform faults*. Although two types of movement take place along the line of the mid-ocean ridge it can be regarded as a single type of plate boundary.

Where two sections of plate are moving towards each other a different set of conditions exists. In this situation one plate margin overrides the other as shown in Fig. 3.14. The overriding plate may be oceanic or continental but, so far as is known, the subducted plate is almost always oceanic. Seismic studies tell us a great deal about the subduction zone. Although the main regional stresses are compressional, several types of stress system operate locally. Firstly, as the plate bends down into the trench some tensional faulting (w) occurs on its upper surface. Secondly there is some strike-slip faulting (x) where the subducted plate slides under the overriding one. Thirdly, the occurrence of tensional faulting (y) indicates that as the plate descends it breaks into segments. Fourthly, the separated segments appear to have internal compressional stresses which produce faulting (z). Finally the zone of seismic disturbance finishes at a depth of about 700 km which suggests that below this the plate has been re-melted and re-absorbed into the mantle.

Plate movement

The geometry of plate motion is that of movement around the surface of a sphere. When two plates move away from each other they are essentially rotating in opposite directions about an axis. This is known as the *axis of rotation*. Fig. 3.15 shows

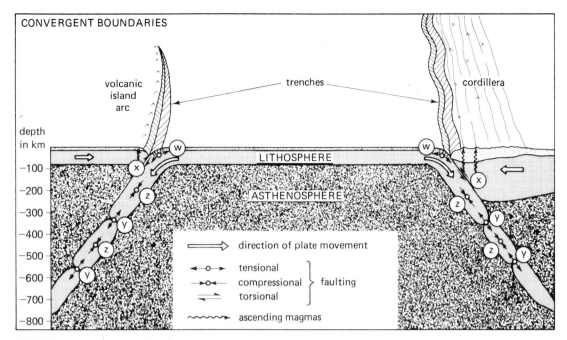

Fig. 3.14 Plate tectonic model of subduction zones

where an imaginary continent has split into two new continents, A and B. The mid-ocean ridge marks the line of the rift and of spreading. It forms the median line between A and B and is composed of two elements; ridge segments and transform faults. The ridge segments are roughly parallel to lines of 'longitude' drawn through the axis of rotation. The transform faults line up with lines of 'latitude' around the axis. They are parallel to the direction of plate movement due to growth. Movement along transform faults then maintains the position and orientation of the ridge. It can be seen that rates of spreading close to the poles of the axis of rotation (P) will be slow relative to those near the 'equator' of movement (Q). Rates of spreading may vary from about 2 cm to about 6 cm per year, per side of the ridge.

Driving mechanisms

It is still not very clear what the forces are which cause plates to move around the Earth's surface. Many mechanisms have been suggested. The earliest theories proposed that the crust was carried along on great convection cells in the mantle (Fig. 3.16a). When the internal structure of the Earth was better known, however, it was realized that flow would be restricted to the asthenosphere (Fig. 3.16b). This meant that convection cells would have to be two hundred times as wide as they were deep. Cells this shape could not exist; they would break up. Besides this Fig. 3.10 shows that there is not a simple 1:1 relationship between ridges and trenches, so the system cannot be driven by a simple system of large convection cells. It has been suggested that plates are buoyed up by numerous small convection cells in the mantle (geothermal turbulence) with a sort of hovercraft effect (Fig. 3.16c). The weight of a continent on one margin would then cause the plate to move in that direction. This, though, would not explain all the Earth's complex plate movements. Another theory is that, as magma is injected along the line of the mid-ocean ridge, it absorbs water, expands and pushes the plates apart (Fig. 3.16d). The structures of the plates, however, are more compatible with a pulling than a pushing mechanism.

Some recent theories suggest that the plates are simply the upper surfaces of convective flow movements in the mantle. These movements are not simple cells, but a more irregular flow pattern. As the material of the asthenosphere wells up at the ocean ridges it cools and solidifies (like the skin, or crust that forms on molten candle wax as it cools). As the flows diverge the upper parts of them will be cool and therefore solid. Furthermore, as they move away from the ridge the plates become progressively cooler and so thicker. Since they are cooler and therefore denser than the parent material on which they are being supported they are in an

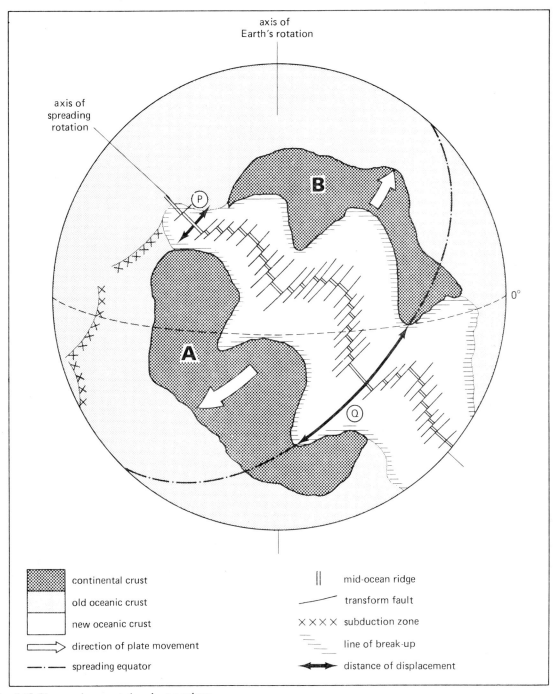

Fig. 3.15 Plate motion as rotation about a sphere

extremely unstable situation. Stability is restored when the plate finally breaks and one edge plunges back down into the mantle and is re-absorbed. The weight of the sinking edge may pull the trailing plate behind it and thus provide a further driving mechanism. The actual situation may be that a combination of these mechanisms causes plate movement but as yet nobody is really in a position to say definitely what the driving force is.

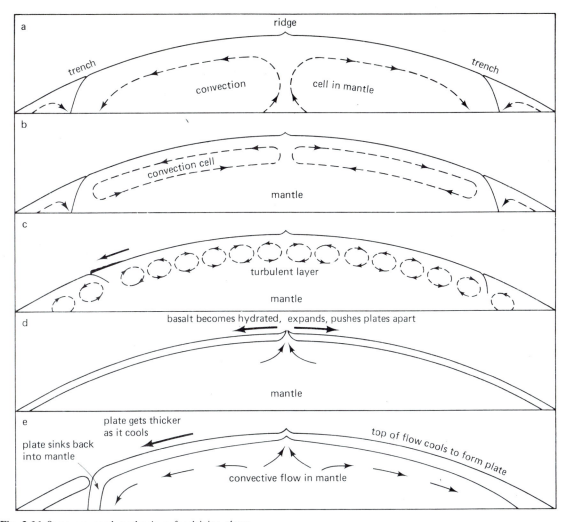

Fig. 3.16 Some proposed mechanisms for driving plates

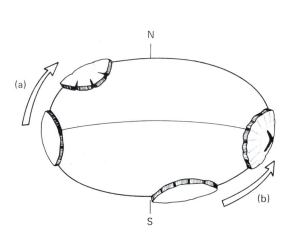

Plates under stress

Plates are relatively rigid and strong, but under stress they may undergo plastic deformation (folding) or brittle deformation (faulting). As mentioned on page 36 the stresses may be tensional, compressional or torsional. Situations of stress may arise at plate margins where adjacent plates interact. Both margins and interiors may be distorted by membrane tectonics as explained in Fig. 3.17. Intense local heating by plumes in the mantle may also create stresses, as explained in Chapter 4.

Fig. 3.17 Stress in plates due to membrane tectonics. (a) Because of the lesser curvature of the Earth near the poles, plates moving from the equator towards the poles have their centre parts in compression and their margins in extension. (b) Motion towards the equator causes the interior to be extended and the margins to be compressed

4

Plates under tension (divergent plates)

Continental breakup

Causes of tension

Those sections of plates which are capped by continental crust will survive considerably longer than those which are capped by oceanic crust. For this reason the majority of features which have been produced on plates subjected to tension are found today in continental crust. The causes of the tension can be surmised by observation of the effects. Two main schools of thought have evolved regarding these causes. One of these relates the stress to divergent mantle flows in the atmosphere; the other relates it to mantle plumes.

The first of these theories maintains that where two convective flows in the mantle are divergent (i.e. they flow away from each other) the viscous drag which they exert on the overlying plate will cause it to be torn apart, in the manner shown in Fig. 4.1.

The second theory is that tensional stress is caused by *plumes*, or 'hot spots', in the mantle. These plumes are localized regions of the mantle where heat flow is well above the surrounding

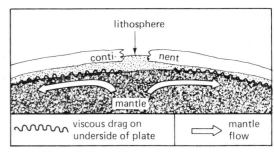

Fig. 4.1 Plate split by mantle currents

average. The causes of these hot spots are not well understood but their effect can be to upwarp, or dome, the overlying crust. Experiments made by Hans Cloos in 1939 show how doming can cause tension. Plate 4.1 shows an experiment made by Cloos in which he domed clay by placing it on top of a slowly inflating hot water bottle. As the clay arched up the tension resulted in a pattern of normal faulting, or rifting, in the clay. This may seem to be rather a simplistic analogy, but Cloos's other experiments, together with his observations of some of the world's major natural rift valleys, do lend support to the 'hot spot' school of thought.

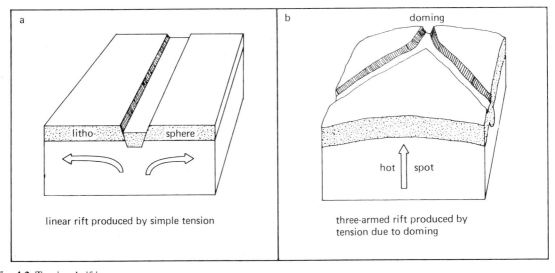

Fig. 4.2 Tensional rifting

Rift Valleys

In his experiments with clay, Cloos was, amongst other things, attempting to reproduce the natural fault patterns occurring in the Rhine Rift Valley. When he applied tension to the clay by pulling the two sides apart he produced a linear rift pattern as represented in Fig. 4.2a. When he caused the clay to be domed over a hot water bottle he produced a three-armed rifting pattern as represented in Fig. 4.2b. This latter effect met his requirements for explaining the Rhine Rift on two counts. Firstly, it explained the plan of the rift, which is three-armed as shown on the sketch map in Fig. 4.3. (The Upper Rhine and Lower Rhine arms are both larger than the Hesse arm.) A comparison of the map with the model shown in plate 4.1 indicates, also, how rifting is not a simple displacement but involves movement along a number of related fault planes. Secondly, Cloos's work provided a possible explanation for the formation of *horsts*, or crustal block mountains, on either side of the rift. (These comprise the Vosges to the west of the rift and the Black Forest Mountains to the East.)

Horsts are slabs of crust left upstanding by the subsidence of rift valley floors. Fig. 4.4a shows a simple tensional explanation for their origin. The model for their formation depicted in Fig. 4.4b, however, may come closer to reality. In nature many horsts take the form more of tilted blocks than of plateaus, which means that their formation is more likely to be due to tension caused by upwarping than by a simple pulling-apart action. Elongate slabs of crust which are downthrown by rifting are known as *graben*. These may comprise

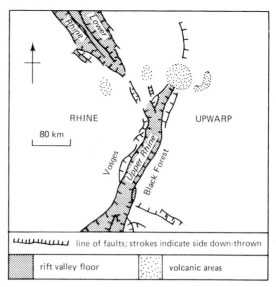

Fig. 4.3 The Rhine Rift Valley

the main blocks of rift valley floors or the flakes involved in the complex faulting on the flanks of rift valleys.

The 'Rheingraben' is only one of many rift valley systems exhibiting a threefold symmetry. An excellent example on a much larger scale is the Red Sea rift system, which shows a similar pattern (Fig. 4.5). In this instance the three arms comprise the Red Sea Rift, the Aden Rift and the Abbyssinian Rift. As in the case of the Hesse Rift (Fig. 4.3) this last rift is smaller than the other two. The Red Sea and Aden rifts have developed to the point where Arabia and Africa have been torn completely apart. Crustal separation has occurred.

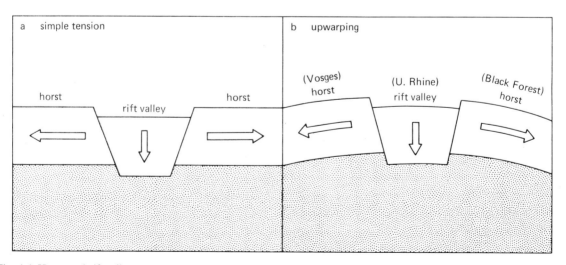

Fig. 4.4 Horsts and rift valleys

Plate 4.1 Experimental production of a rift valley by slow upheaval of layers of moist clay. Note that the surface width of the rift valley is of the same order as the total thickness of the clay 'crust' (*Hans Cloos*, 1939)

Crustal separation

On the above evidence it may be suggested that heating by mantle plumes, causing doming and rifting, may be the mechanism which causes the splitting of plates that can tear continents apart. The locations of plumes in the mantle are detected by the presence of 'hot spots' in the crust. These hot spots have three main identifying characteristics: as mentioned on page 42 they are regions where the flow of geothermal heat up through the crust from the mantle is considerably higher than the average; they are commonly sites of vulcanicity and the lavas produced have a characteristic composition, being rich in the alkali metals (lithium, sodium, potassium, etc.); they can cause upwarping of the crust forming domes up to 200 km across.

The locations of plumes are believed to have remained relatively static within the mantle. Continents which hold stationary positions over plumes are considerably affected by the heat, while those travelling across plumes remain relatively unaffected. This has been compared to drawing a sheet of paper across a glowing cigarette-tip—as long as the paper moves steadily it will only be scorched on the underside, but as soon as it comes to rest it will be burnt right through. It is thought that Africa has held a static position for the past 30 million years. During this period alkali basalt lavas, characteristic of plumes, have accumulated at certain centres, superimposed one on top of the other. If the plate had moved one would expect to find the eruptions strung out in lines marking the plate's progress across the plumes. About a sixth of the world's known plumes underlie the African continent. The broad topography of Africa is one of *swells and basins*. These swells may well be the product of upwarping over plumes, with the basins being the intervening lowlands (Fig. 4.6).

Comparing Figs. 4.5 and 4.6 one can see how the crests of upwarps, or swells, in East Africa have fractured, forming the East African Rift Valley. This is something of a misnomer since in reality it comprises a whole complex of rift valleys, which form a system running down the east of the continent. Volcanic activity in these rift valleys typically produces alkali basalt lavas. The Abyssinian Rift, although it appears to be a continuation of the East African Rift, is part of the Red Sea rift system. This too appears to be due to faulting along the crest of an upwarp. This all ties in with the theory which suggests than when a continent comes to rest over a plume, upwarping, rifting and continental breakup may ensue. More evidence that this may

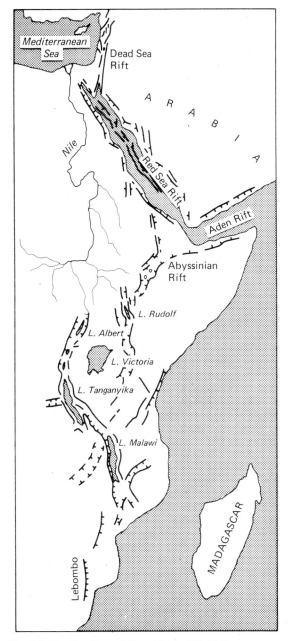

Fig. 4.5 The Red Sea and East African Rift Valleys

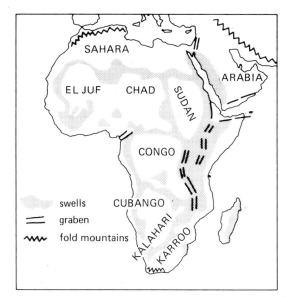

Fig. 4.6 The swell and basin topography of Africa

be the case can be found by studying the three-armed pattern of rifting described in relation to the Rhine and Red Sea rift valleys.

In both these instances one finds the occurrence of two major arms and one minor arm. In the case of the Red Sea system the major rifts, the Red Sea Rift and Aden Rift, go to form an embryonic ocean whilst the minor arm, the Abyssinian Rift, may be termed a failed arm, in the sense that it has failed to develop into a new ocean. How this process can split continents to form new oceans is illustrated in Fig. 4.7. The inset shows the Red Sea situation as described. The main, refit, map shows how the Atlantic Ocean has developed from a series of rifts which have merged into a chain. In most cases two arms of a system are included into the ocean, whilst the third, failed, arm extends as a rift valley into an adjacent continent.

It is thus no coincidence that, of those plumes which underlie the ocean floor, a significant proportion lie under, or close to, mid-ocean ridges. These plumes were possibly those which initiated rifting to form the oceans in the first place. Considering this evidence it does not seem unreasonable to accept that plumes in the mantle are a significant cause in the splitting of plates and the breakup of continents.

The margins of new continents

Plateau Basalts

The lavas erupted from above plumes are alkali basalts. Once splitting has occurred, however, the basalts produced along most of the length of the fissure (i.e. all except those points which directly overlie plumes) are of a different composition. These are poorer in potassium and the alkali metals but are richer in silica. They are called *tholeiitic basalts*. Besides being produced along the mid-ocean ridges these tholeiitic basalts are found in

Fig. 4.7 The splitting open of the Atlantic by three-armed rifts

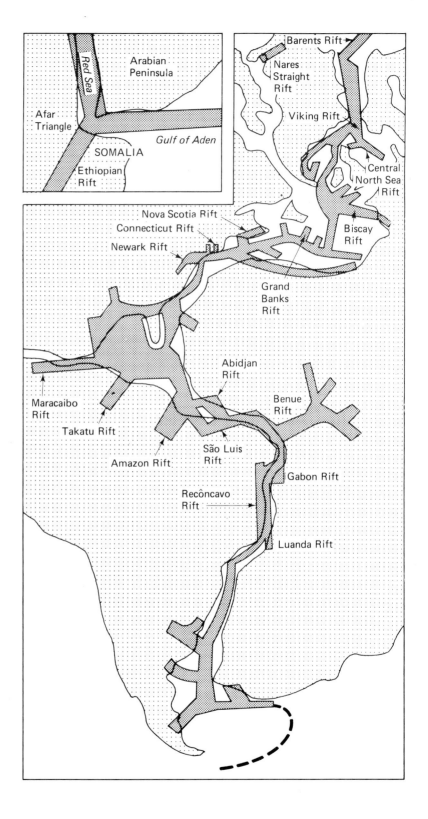

Plate 4.2 Black ribbons of fresh basalt in the Erta'Ale mountains mark zones where molten rock has poured out of fissures in the floor of the Afar triangle. The basalt is chemically similar to the magmas that have welled up from the rifts in the earth's mid-oceanic ridges.

great sheets on the margins of some continents. Fig. 4.8 shows their occurrence on a southern continents refit map. The reason why they lie on continental margins is explained by their mode of origin. When continents start to tear apart, great volumes of tholeiitic basalt well up through the fissures and flood wide areas. Once separation has occurred these sheets of lava will be situated at the new continental margins. Plate 4.2 shows outpouring of tholeiitic basalt from a fissure on the margin of the Red Sea Rift, in an area called the Afar Triangle. (This is a triangular area of overlap found when Africa and Arabia are refitted together. It is believed to be an area of oceanic crust which lies above sea level.)

Basalt is relatively resistant to erosion. The rock which basalt lava-flows overlie may not be quite so resistant. If this is the case, when the level of the land around the flow is reduced by erosion, the basalt may act as a caprock and the area protected

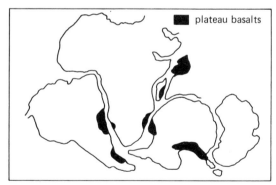

Fig. 4.8 Plateau basalts of the southern continents

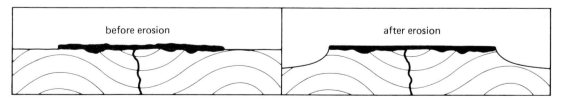

Fig. 4.9 The development of a basalt plateau

This photograph, taken from a height of 130 nautical miles by *Apollo 7*, shows where India (left) collided with Asia (right), throwing up the Himalayas.

by the flow may be left upstanding as a plateau (Fig. 4.9). For this reason continental tholeiitic basalts are commonly known as *plateau basalts*. An excellent example of a plateau formed in this manner is the Deccan Plateau of West India. The best example in the British Isles is the Antrim Plateau, in northeast Ireland (Fig. 4.10). This was formed some 50–55 million years ago by movements related to the opening up of the North Atlantic. The fact that this lava was extruded sub-aerially, rather than underwater, is testified to by the hexagonal columns of the Giant's Causeway, structures which could not have formed in submarine conditions. The finding of charred remains of trees and peat between lava flows here indicates that extrusion of plateau basalts may be episodic, rather than continuous.

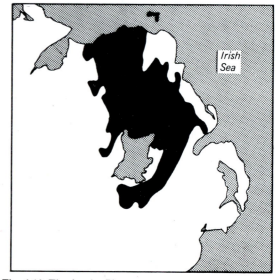

Fig. 4.10 The Antrim Plateau

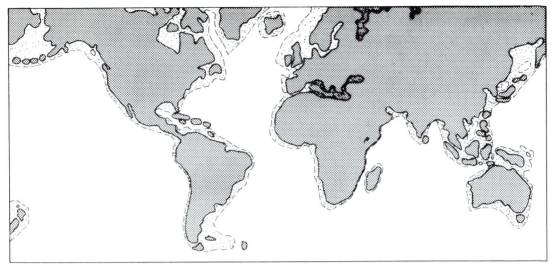

Fig. 4.11 Continental shelves of the world

Continental shelves

Those areas which are part of continents yet which lie below present sea-level are known as *continental shelves* and *continental slopes*. Since sea levels can drop up to 150 m during periods of glaciation, such areas may, however, be substantially uncovered for long periods of time. This may lead one to suppose that continental shelves are superficial topographic features, which is not the case. Fig. 4.12 shows how the width of continental shelves may vary considerably. In general those on continental margins which are moving apart (e.g. Europe and North America) are broader than those on margins which are drawing together (e.g. North America and Asia). This is to say that continental shelves are wider in areas of tension than compression and their structure, therefore, should be related to plate movement. An examination of continental margins under tensional conditions shows how continental shelves only make up part of their structure.

A cross-sectional profile through such a margin is represented in Fig. 4.12. The jagged line indicates the line of suture where the original split took place and where the new oceanic plate is welded on. As shown, the continental plate is thinner at the margin where it was torn apart. The effect of this is to produce several distinct topographic elements in the profile. The shallow sea-floor fringing the land mass is the continental shelf proper. This may be overlain by an accumulation of sediments eroded from the adjacent land, though in some instances

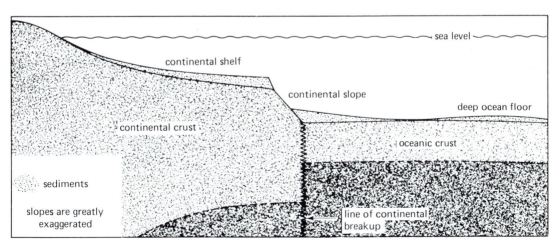

Fig. 4.12 Cross-section of a continental shelf

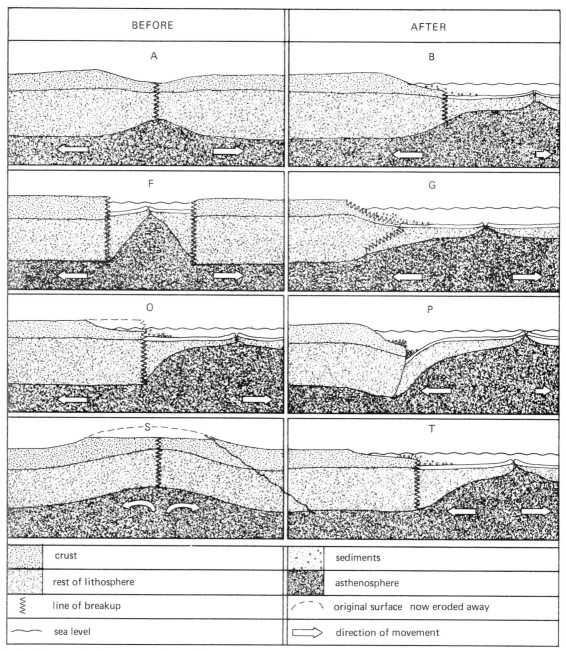

Fig. 4.13 Four mechanisms suggested to explain the origin of continental shelves

all but the most recent sediments may have been scoured off by glaciation or other processes. Further offshore the sea-floor drops more steeply an average of 2 km to the *deep ocean floor*. This drop is the *continental slope*. Whereas the average gradient of the continental shelves is 0° 17′, the gradient of the continental slope averages 4°. The deep ocean floor is the surface of the new oceanic crust generated to seal the rift which was formed by the separation of plates. Sediments may also accumulate on the deep ocean floor at the base of the continental slope.

It used to be thought that continental shelves were simply the product of coastal erosion and deposition. Now it is known that they are part of the marginal geological structure of continents. There is, however, no consensus of opinion among geologists as to how these structures developed.

Four different theories put forward to explain them are shown in Fig. 4.13. Diagrams B, G, P and T on the right show end products which are essentially similar. Diagrams A, F, O and S on the left show the mechanisms invoked for their formation and they can be seen to vary considerably.

Theory AB proposes that before the plate breaks in two it is drawn out into a thin neck. After this *ductile necking* the plate fractures and tears apart, leaving a 'neck' on each new continental margin as a continental shelf.

Theory FG proposes that the initial break is a clean one, but after this gravity acts on the unsupported broken edge, causing it to bulge outwards. This slumped-out margin then forms the continental shelf.

Although sedimentation is depicted in the other theories, only in OP is it involved as part of the mechanism. In OP it is suggested that a clean break occurs and then erosion wears down the margin. The eroded sediments are deposited on the deep ocean floor which warps down under their weight to give the situation shown in P.

Theory ST suggests that doming occurs over a hot spot and the top of the dome thus produced is worn down by erosion. When breakup takes place, therefore, the marginal thickness of the crust will have been reduced and will subside (due to isostasy), producing continental shelves as shown in T.

Geosynclinal sedimentation

The term *geosyncline* is used to describe a downfolded trough in the crust of the Earth. It was thought that sediments accumulating in this trough would subsequently become the rocks of the fold mountain belts of the world (Fig. 4.14). This model has now been discarded as it is incompatible with present knowledge of geological structure. A far better model can be constructed on the basis of the continental marginal structure described in the previous section.

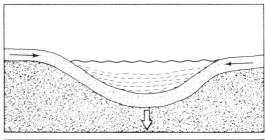

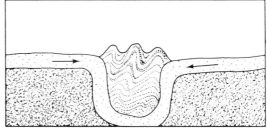

Fig. 4.14 The original concept of a geosyncline

Both the continental shelves and the bases of continental slopes can act as sites for the accumulation of great wedges of sediment eroded from adjacent land masses. They also provide a number of diverse sedimentary environments to accommodate the varied lithologies of marine sedimentary rocks, as shown in Fig. 4.15. By means of the lithology and sedimentary structures it is often possible to identify the environment in which a sedimentary rock has been deposited.

The sedimentary environments found on continental shelves may be loosely grouped as coastal environments. Such sedimentary features as beaches, sand-bars, estuaries, lagoons and deltas may be produced in coastal environments. Given suitable conditions of light and temperature for the growth of coral, there may be offshore development of reefs. The shallow sea-floor of the shelf proper provides a major sedimentary environment. There all grades of eroded rock debris may be deposited, ranging from pebbles to mud and producing a corresponding range of lithologies from

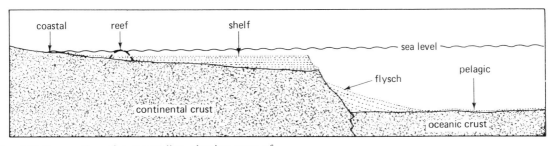

Fig. 4.15 Cross-section of a geosyncline, showing types of sediment

conglomerate through to shale. The various grades of sediment are usually differentiated and laid down in thin layers, or *beds*. Since the shallow shelf sea provides a habitat for a considerable range of flora and fauna, their remains are abundant. Fossils usually occur in the rocks. More importantly, broken down calcareous skeletal material forms the basis for the formation of limestones, particularly in tropical waters.

As sediments are deposited on the edge of the continental shelf they reach a point where they become banked up so steeply as to be unstable. Once this point is reached masses of sediment may break away, perhaps triggered by earthquake shocks, and spill down the continental slope (Fig. 4.16). The sediment becomes mixed with sea-water and the resulting suspension flows down the slope at speeds of up to fifty miles per hour. This flow is called a *turbidity current*. All grades of sediment may be incorporated into the load of the turbidity current. When it slows down at the base of the slope it will drop its load. The coarsest, heaviest material will be dropped first and then the finer, lighter sediment will be deposited on top of that. Thus the sediments at the base of the slope will be graded as shown in Fig. 4.16. Ancient layers of alternating sandstone and shale formed in this manner and now exposed in the Alps are known by the Swiss dialect term *flysch*.

The long prisms of sediment which collect in these geosynclines are important since they provide the building material for chains of fold mountains, as described in Chapter 5.

The expanding ocean floor

Ridge Features

The only major site where a mid-ocean ridge appears above sea level is Iceland. Fig. 3.4 illustrates how Iceland grows outwards from the centre, being constantly pulled apart while lavas well up from below to seal the fissures. Though on the centrefold map Iceland might appear to be a block of continental crust it is in fact a tholeiitic basalt plateau of the sort described on page 45. In its case the extrusions occurred after separation had taken place and so they formed a platform straddling the mid-Atlantic Ridge. A rift valley, the Central Icelandic Depression, lies down the centre of the island, roughly coinciding with the belt of most recent volcanic activity. Plate 4.3 shows the graben related to this rifting. This central rift is seen in

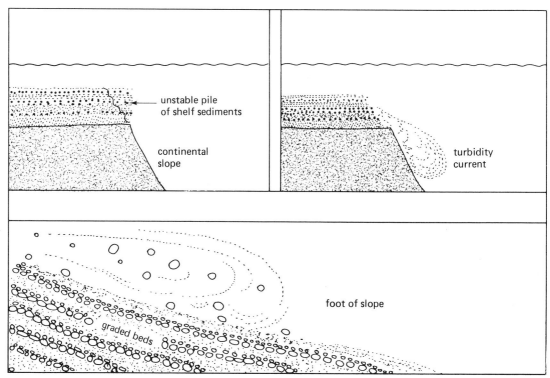

Fig. 4.16 Turbidity current and Flysch formation

Plate 4.3 Aerial view of the Thingvellir rift valley in Iceland showing a graben boundered by normal faults. Rifting of this type develops as sea-floor spreading causes the island to be pulled apart. (Courtesy of Sigurdur Thorarinsson.)

most mid-ocean ridges, though it is best displayed in the mid-Atlantic Ridge.

The Red Sea also has a well-defined central rift valley, where water temperatures reach 56°C and salinities of 256 parts per 1000 have been recorded. (35 parts per 1000 is the norm.) These high temperatures could be explained by the high geothermal heat flow and the underwater extrusion of lavas. Where basalt lava-flows have been photographed on the flanks of the mid-Atlantic Ridge (at depths of over 2500 m) their surface is seen to have a characteristic pillow-like appearance. This is caused by rapid cooling underwater (cf. hexagonal columns in sub-aerially cooled flows). Being slightly poorer in potassium and richer in aluminium than land basalts these are sometimes described as oceanic tholeiites.

The other dominant features of mid-oceanic ridges are the great transform faults, which have already been described on page 26.

Guyots and atolls

Guyots are flat-topped volcanic cones found on the deep-ocean floor. One popular theory for their

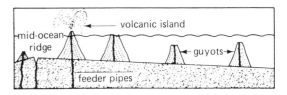

Fig. 4.17 The formation of guyots

origin is that they were originally volcanic islands extruded at mid-ocean ridges or over plumes (see page 54). On being carried away from the ridge on the moving oceanic plate they would have moved into deeper water. The tops of the cones could have been truncated by wave erosion so that, by the time they became submerged, the guyots would have acquired their characteristic flat-topped form (Fig. 4.17). An alternative theory, that they are built up from layer upon layer of submarine volcanic ejecta, has been suggested by Haroun Tazieff, the famous French volcanologist. His theory is based on studies of the guyot-shaped cone shown in Plate 4.4, which lies in the Afar triangle.

Atolls are ring-shaped coral islands. Fig. 4.18

shows how in their formation they may be related to guyots. They may start their lives as reefs fringing volcanic islands. As the islands sink below the waves the coral may grow upwards to form the ring-shaped reef called an atoll (Plate 4.5). (This theory was originally suggested by Charles Darwin.)

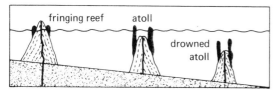

Fig. 4.18 The formation of atolls

Plume volcanic islands

Some volcanic islands on the deep ocean floor are not related to mid-ocean ridges (nor to Benioff Zones as are those described in Chapter 5). These lie in plate interiors rather than at the margins. The Tristan da Cunha group in the South Atlantic Ocean and the Hawaiian group in the Pacific Ocean are examples. Both these groups are believed to have been produced by plumes in the mantle, but under different circumstances.

In the case of the Tristan da Cunha group the plate they lie on (the African Plate) is static. As a result, alkali basalt lavas have been erupted at this one site overlying the hot spot for the past 18 million years. Previous to this the plate had been moving over the plume which accounts for the chain of volcanic outpourings on the sea floor called the Walvis Ridge.

The Hawaiian chain of islands marks the progress of an oceanic plate across a plume. The process of formation is illustrated in Fig. 4.19. The cone directly overlying the plume will be active. As the plate moves on and carries the cone away from the plume it will become inactive. As early as 1828 James Dana, an American geologist, deduced from their state of erosion that, as one traced the chain to the northwest, the islands became progressively older. Bearing in mind the analogy of the sheet of paper being drawn across the cigarette tip, one can see how these volcanic chains can be used to retrace the movement of the plate across the plume. Fig. 4.20 shows the trace left by the Hawaiian plume on the Pacific Plate. The active volcanoes of Pitcairn Island and the MacDonald Seamount may possibly suggest the presence of two more plumes. The kinks in the island chains indicate that about 40 million years ago the Pacific Plate changed its direction of movement.

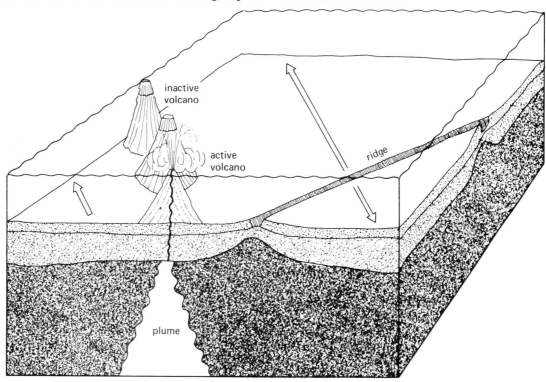

Fig. 4.19 The origin of a plume volcanic island chain

Fig. 4.20 Plume volcanic island chains of the Pacific Ocean floor

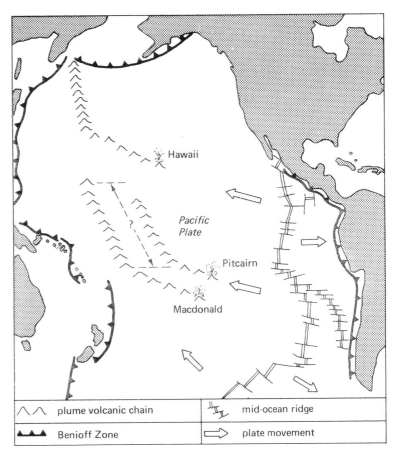

Plate 4.4 Flat-topped volcano, Mount Asmara is composed of shards of volcanic glass such as are formed during underwater volcanic explosions. It resembles the numerous guyots, or submerged oceanic mountains, whose level summits are usually attributed to wave erosion Because Mount Asmara was formed under water, it may be that a flat top is instead a feature common to all such volcanoes.

Pelagic sediments

Where the deep ocean floor is being generated, at the mid-ocean ridge, it is almost entirely bare of sediments. For up to 100 km from the ridge crest the depth of sediment may hardly be measurable. Where sediments do accumulate they contain very little *terrigenous* (land derived) elements. Their main components are chemical and organic. These characteristic deep ocean-floor deposits are called *pelagic sediments*.

The dominant terrigenous sediments found on the ocean floor are *red clays*. These are derived largely from wind-borne dust which can be carried far out to sea. They may be augmented by volcanic ash and material dropped from melting icebergs. Typically they are red to chocolate coloured, being highly ferruginous. Manganese is also present in high concentrations, and is sometimes, with iron, found as oxide-hydroxide nodules.

The bulk of the organic content of pelagic sediments is composed of the skeletons of *plankton*. These are tiny primitive sea organisms which live in the illuminated zone of the oceans, near the surface. As they die their skeletons, in millions, transported in the form of faecal pellets, form a continuous 'rain' down to the sea floor. The bulk of these skeletons are composed of calcium carbonate and when these land on the sea floor they form a *calcareous ooze*. Below a certain depth, however, these are dissolved and the smaller fraction of skeletal material, that composed of silica, forms the dominant sediment. These silicious skeletons are largely those of *radiolaria* and *diatoms* and the sediment, when it lithifies, forms bands of *chert*.

5

Plates under compression (convergent plates)

Configurations of convergence

As any two plates move away from one another their movement takes the form of rotation about the surface of a sphere. It is a corollary of this that, simultaneously, these same two plates are moving towards each other on the opposite side of the sphere. Thus it is that America and Eurasia, whilst diverging at the Atlantic Ocean, are converging at the Pacific Ocean. At the same rate that the Atlantic is expanding, the Pacific is being consumed. It should be noted that it is the ocean floor which is being consumed and not either of the two continents. This is because, as mentioned in Chapter 4, those sections of plate which are capped by continental crust, being less dense and therefore more buoyant, are infinitely more durable than those which are capped by oceanic crust.

At sites where plates converge on one another the main stresses which occur are compressional and so the dominant features produced are compressional features. The nature of these features depends to a considerable extent on the type of crust capping the plates. Oceanic crust is subductable, which is to say that it is capable of being thrust down into, and re-absorbed by, the mantle. Continental crust is not subductable, and neither is the crust which is produced under volcanic island arcs. When oceanic crust is involved in convergence it may be subducted, with compressional features being produced at the site of subduction. An alternative situation, however, arises where both of the convergent plates are non-subductable, for there a collision will ensue. Compressional features will be produced, but with the two involved plates locked together convergence will virtually cease at that site. Plate movement about that pole of rotation must then either virtually cease or subduction must

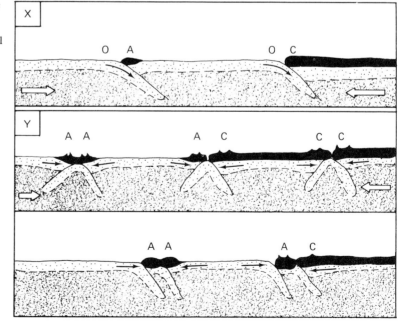

Fig. 5.1 Configurations of convergence
 X: subductive convergence
 Y: collision convergence
 (O = oceanic plate; C = continental plate; A = island arc system)

commence at some other site. (In some instances plate movement does continue, but in a new direction and about a new pole of rotation.) Figure 5.1 shows seven possible configurations of convergence. The two configurations shown in Fig. 5.1X are subductive. The five shown in Fig. 5.1Y are all collision configurations, where buoyant plate margins have crashed together and subduction has ceased. All these convergent boundaries have, however, one element in common; they are all sites for the production of compressional tectonic features.

The most significant process which takes place along compressional plate boundaries is that of *orogenesis*, or mountain building. This occurs where the great wedges of sediments lain down in geosynclines are compressed by folding and faulting to form chains of mountains such as the Andes, the Alps and the Himalayas. As will be shown, orogenesis can take place in differing compressional situations, so producing varying types of mountain range.

In Chapter 4 it was demonstrated how normal faulting (rifting) is produced in plates subjected to tension. Reverse (thrust) faulting is produced by compression. The third type of faulting (see Fig 3.12) is a response to torsional stress. This torsion can be created where plates converge obliquely and it may produce strike-slip (wrench) faults, so these too will be dealt with as features of convergent plate margins.

Subductive convergence

Oceanic subduction zones

In Chapter 3 it was described how, when one oceanic plate is forced down under another oceanic plate, an arcuate subduction zone is formed. The reason for this arcuate form of fracture becomes apparent if one tries to dent or push a hole in a table-tennis ball. The bend or break takes place along a curved line. This is because a thin spherical shell, which can bend but not stretch, can only bend or break in on itself along an arcuate line (see Fig. 3.14). Where a plate under compression is free from restraints such as an overriding continent it will, when it fractures, form a subduction zone which adopts this arcuate pattern.

The structure of a subducted plate is shown in Fig. 3.14. The descending plate has a considerable influence on the overriding plate, notably the creation of an arc of volcanic islands set back about 100 km from its rim. The lavas of these island arc volcanoes are dominantly composed of *andesite*, which contains about 15 per cent more silica and about 3 times the amount of K_2O (by weight) than the ordinary tholeiitic basalt of the ocean floor. Traditionally the origin of these andesites has been ascribed to a melt of ocean floor tholeiite and wet sediments which had been carried down the Benioff Zone. More recent studies, however, indicate that andesites are probably the product of partial melting of the upper mantle under conditions of high pressure and low temperatures (caused by the cold descending slab) and in the presence of water (derived from subducted wet sediments). The composition of the andesites in island arc volcanoes varies in proportion to distance from the trench (i.e. in proportion to the depth of the Benioff Zone: see Fig. 5.2). The heat required to cause re-melting is produced from friction generated by the subducted plate grating along the underside of the overriding plate.

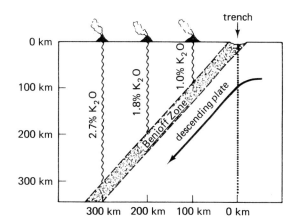

Fig. 5.2 Compositional variations of andesitic lavas related to the depth of the Benioff Zone

The descending slab of the subducted plate, as it grinds under the overriding plate and down into the mantle, is much cooler than the material into which it is plunging. This creates conditions of higher pressures and lower temperatures than the surrounding geological environment, and so tends to alter, or *metamorphose* the local rocks. This type of alteration is an example of *dynamic metamorphism*. In this particular situation, due to the colour of some of the newly generated minerals, it is sometimes known as *blueschist metamorphism*. The rim of the overriding oceanic plate may therefore constitute a dynamic, or blueschist, metamorphic zone.

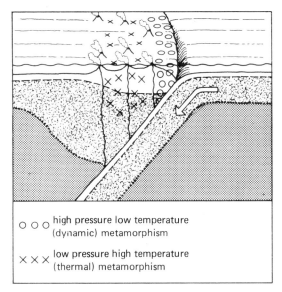

Fig. 5.3 The formation of paired metamorphic belts over subduction zones

Further in from the rim of the plate a totally different sort of geological environment exists. There are, due to the ascending magmas from the Benioff Zone, very high temperature conditions. Drag, produced in the wake of the subducted slab as it sinks into the mantle, may actually cause conditions of tension. This high-temperature, low-pressure environment causes *thermal metamorphism*, which is to say alteration of the local rocks by heat. In this situation the metamorphosed minerals may take on a greenish hue, and hence the process may be called *greenschist metamorphism*. Thus a second, thermal metamorphic belt may be produced, parallel to the first, dynamic one. These paired metamorphic belts may be used as criteria for establishing the existence of former subduction zones.

Fig. 3.6 shows how, in the Western Pacific, a system of island arcs lying off the continental margin has developed. The partially enclosed seas, trapped between arcs and either continental or other arcs are known as *retro-arc basins* or, alternatively, as *marginal basins*. The basins act as settling traps to catch sediments eroded down from the adjacent continents and volcanic islands. Retro-arc basins can, therefore, act as geosynclines, accumulating great masses of sediment (Figs. 5.4 and 5.7). These basins are also notable for another reason. They may act as sites of *crustal extension* at rates which have been calculated as being up to 1 cm/yr. The crustal growth is not along a single fracture, as at a mid-ocean ridge but, instead, as many intrusions into a complex swarm of fractures on the basin floor. These fractures could be produced as a response to tension caused by the oceanward migration of the island arc as the subducted plate sinks into the asthenosphere (Fig. 5.4).

Continental marginal subduction zones

When an ocean is expanding there is no need for subduction, since the increased area of plate is accommodated by sideways displacement of the adjacent continent (Fig. 5.5a). If the continent stops moving, however, as shown in Fig. 5.5b, the thin oceanic plate will fracture and the increased area of crust will be accommodated by subduction down into the mantle. The fractures will be offshore and assume the ideal arcuate pattern. Should the ocean start to contract, with the continental plate advancing on the oceanic plate (Fig. 5.5c),

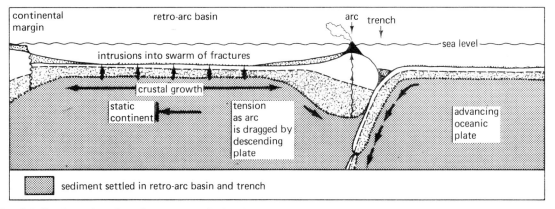

Fig. 5.4 Sketch section through a retro-arc basin showing crustal extension and deposition of sediments

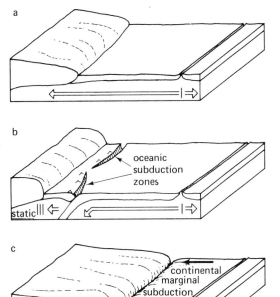

Fig. 5.5 The formation of subduction zones:
(a) Continental crust and oceanic crust moving together: no subduction; (b) Oceanic plate advancing on static continental plate: formation of offshore arcuate subduction zones; (c) Continental plate advancing on and overriding oceanic plate: formation of continental marginal subduction zone

this arcuate form will not be able to develop, and the subduction zone will instead develop along the rim of the continental plate as it overrides the ocean floor. This is why a *continental marginal subduction zone* will form in situations where a continental plate is advancing across an oceanic one.

During the period that an ocean is expanding, the trailing edge, or margin, of the adjacent continent is an area of geosynclinal sedimentation (page 51). When the direction of plate movement is reversed the old trailing edge of the continent becomes the new leading edge, and the site of compressional rather than tensional stress. Those sediments which had been laid down in the marginal geosyncline become compressed between the two converging plates and squeezed up to form a

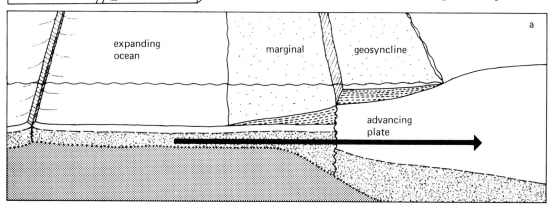

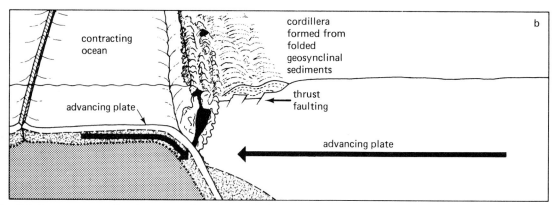

Fig. 5.6 Convergence causing the compression of a marginal geosyncline to form a cordillera. (For more detailed structure see Fig. 5.8.)

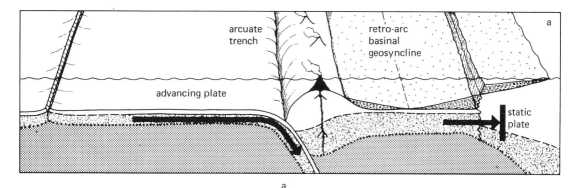

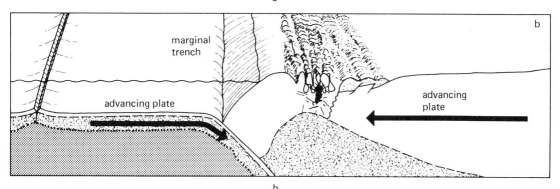

Fig. 5.7 The incorporation of an offshore island arc system into a cordillera. (Example: the Andes.)

cordillera, or mountain chain. This is one instance of orogenesis.

The Andes provide a fine example of a cordillera. The structure of this particular cordillera indicates, however, that its development was slightly more complex than the simple system just described. In its case, as illustrated in Fig. 5.7, the continent remained static for a period, allowing the development of offshore island arcs and retro-arc basinal geosynclines. Subsequently South America started its move westwards across the floor of the Pacific Ocean. Both the island arcs and the geosynclines became incorporated into the new Andean Cordillera and the Peru-Chile marginal trench was created.

Looking at the structure of the geosynclines in Figs. 5.6 and 5.7 one can see that they are each made up of two distinctive elements. The sediments which have accumulated at the base of the continental slope are laid down on oceanic crust, or sima, and are said to be *ensimatic*. The sediments accumulating in a retro-arc basin are also ensimatic. That part of a geosyncline where the sediments are laid down on oceanic crust is called the *eugeosyncline*. The second element of the geosyncline is the *miogeosyncline*, where the sediments are *ensialic*, or deposited on continental crust (sial). As shown in Fig. 5.8 the miogeosyncline comprises the sediments which have accumulated on the continental shelf. The eugeosyncline and the miogeosyncline run parallel to one another and are together known as the *geosynclinal couplet*.

On being subjected to compressional stress these two elements act in rather different manner. The miogeosyncline, being laid down on a strong continental crustal foundation, will only be crumpled into folds, with some thrust faulting extending up through the crust into the sediments (Fig. 5.8). The eugeosyncline, however, has no such support and the effects of compression on it will be much more dramatic. The sediment will be extremely contorted by folding and faulting. Not only this but they may be metamorphosed by the effects of the subduction zone. Close to the zone there will be high-pressure, low-temperature metamorphism. Further away from the zone there will be low-pressure, high-temperature metamorphism. The effects of the latter, the thermal metamorphism, are considerable. It has been calculated that *greywackes*, rock types characteristic of eugeosynclines, can, at relatively low temperatures and in the presence of water, melt to form granitic magmas. Indeed, the hearts of eugeosynclinal fold mountains are often composed of *batholiths*, or *plutons* of granite or a similar rock-type, *granodiorite*. Eugeosynclines are also distinguished by the pres-

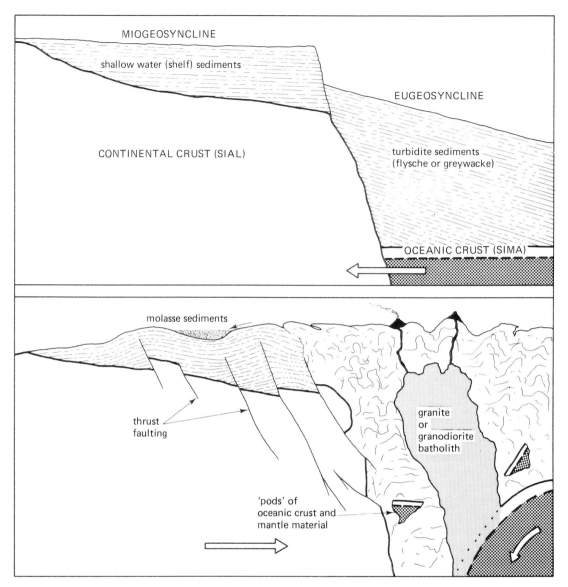

Fig. 5.8 Orogenesis. A marginal geosyncline is compressed into a cordillera. The miogeosynclinal sediments, on a strong sial base, are relatively undeformed. The eugeosynclinal sediments, on weak oceanic crust, crumple up: thermal metamorphism causes remelting producing batholiths and volcanics: fragments of oceanic crust may be incorporated into the cordillera

ence of andesitic volcanoes; andesite being the characteristic lava in areas overlying Benioff Zones. To sum up, the cordillera will be built up from two parallel ranges of mountains; one the miogeosynclinal range, folded but not highly metamorphosed; the other the eugeosynclinal range, highly folded and metamorphosed, with abundant andesitic volcanoes, granitic batholiths and possibly *ophiolites*—fragments of oceanic plate (Fig. 5.8).

Once a new mountain range has been thrown up the elements proceed to wear it down again. Some of the eroded sediments will be transported to the oceanward side of the cordillera and deposited in the sea. The rest will be transported to the continental side and accumulate as fluvial and lacustrine sediments known as *molasse*.

The topmost layer of oceanic crust, that composed of pelagic sediments (see page 56), is not mechanically coupled to the underlying plate. As the subducted plate slides under an island arc or cordillera therefore, the sediments may be skimmed off and accumulate against the back wall of the trench, as illustrated in Fig. 5.9a. There are sites, however, such as the Chile/Peru Trench, where these ocean floor scrapings are nowhere in evidence. Comparing the areas of the Pacific floor west and east of the East Pacific Rise one can see that a considerable area of crust which once lay to the east of the ridge must since have been subducted, so it is necessary to explain the whereabouts of a considerable volume of pelagic sediments. One might suggest that they have been subducted along with the underlying plate, (Fig. 5.9b), but pelagic sediments are not attached to the plate and, having low densities (2.4 gm/cc), they would be unlikely to have sunk into the mantle of their own accord. It is possible that they may have been sucked down in the wake of the descending slab, or that the overriding continent has simply 'bulldozed' its way over them. It should be noted that the presence of wet sediments down the Benioff Zone helps to explain the composition of igneous rocks (i.e. granites and andesites) in the overriding plate.

An alternative explanation for the removal of pelagic scrapings is illustrated in Fig. 5.9c, which shows a situation where convergence is oblique rather than 'head-on'. This creates torsional stress along the junction and the prism of sediments is displaced along the overriding plate margin by *strike-slip (wrench) faulting*. An actual example of this is the northern section of the San Andreas Fault, in California (Fig. 5.10).

The southern section of this fault was produced by the same type of stress, but caused by a different mechanism. The centrefold map shows how part of the East Pacific Rise has been overridden by the North American continent. At the point where the ridge ran under California its divergent movement split off a flake of America to form part of the Baja California Peninsula (Fig. 5.10d). The remainder of the peninsula was broken away by wrench faulting caused by the oblique movement of the Pacific Plate (Figs. 5.10d–e). These faults demonstrate how strike-slip faulting can develop along the margin of obliquely converging plates.

Collision convergence

Continent and island arc collision

In describing the evolution of the Andes in the previous section it was shown how an offshore island arc system was incorporated into the margin of the South American continent (Fig. 5.7). Since neither the island arc system nor the continent was subductable they collided and locked together.

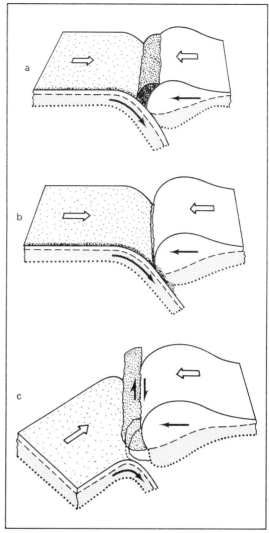

Fig. 5.9 Abyssal sediments and subduction: (a) Sediments skimmed off to accumulate against overriding plate rim; (b) Sediments subducted along with oceanic plate; (c) Sediments accumulate and are displaced by strike-slip faulting caused by oblique convergence

This type of continental growth is known as *accretion*.

The Klamath Mountains in the Cascade Range of the west of the United States are another possible example of this type of event. They contain the remnants of both Silurian and Devonian/Permian island arc systems, together with fragments of ocean crust and mantle. The history of the area may be postulated as: development of an offshore arc system; incorporation of the arcs into the continental margin; development of a new offshore arc system; incorporation of the second arc system

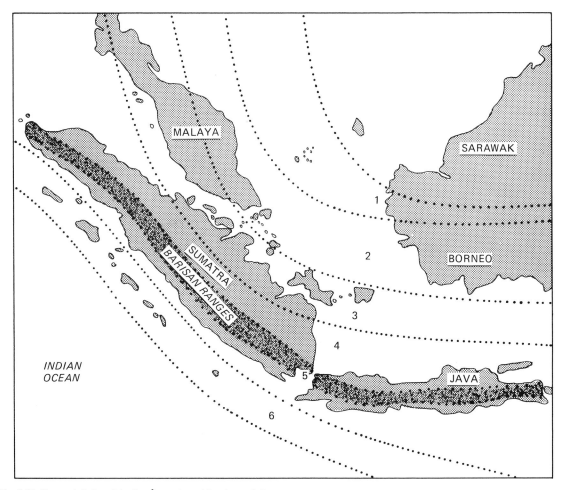

Fig. 5.10 Orogenic belts of the Sundra region of Southeast Asia, numbered in order of formation

by the continent. This type of continental growth by the accretion of island arcs has also been suggested as the mode of formation of the Sunda region in Southeast Asia.

The Sunda region is built up from a sequence of arcuate structural belts which have been pushed up against one another in succession. Fig. 5.10 is a map of the area with the approximate position of the belts marked in. The numbers show the order of formation of the belts. Each belt represents a cycle of mountain-building activity. One theory to explain the structure of the area is that it is compounded from a succession of island arcs and their basins which have been accreted on to the continental margin. As each arc collides with the continent it becomes the new continental margin. Each subsequent collision is therefore between the newly extended continental margin and a new island arc which has developed offshore.

Continents in collision

Referring back to Fig. 2.5 one can see that, on the ancient 'supercontinent' of Pangaea, the majority of mountain chains were marginal, as are the Andes today. Some however, like the Urals, formed barriers across the continental interior. The Himalayas, separating India from Asia, are in a comparable situation at present. Continents involved in collision, since they cannot be subducted, lock together to form newer and larger continents. Intra-continental cordilleras are thrown up along the zones of collision between older and smaller continents. This is another situation of orogenesis, but these cordilleras tend to be much more complex in structure than their marginal counterparts. The reason for this is that there are so many more variable factors involved.

Some of the possible types of suture are illus-

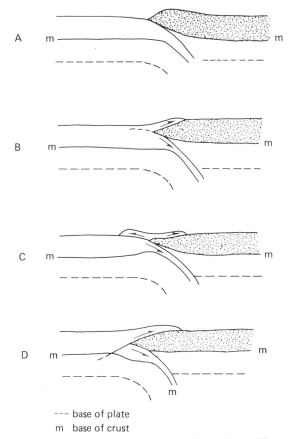

Fig. 5.11 Collisions between continents: (a) simple overridings; (b) flake formation; (c) a detached flake; (d) a deep flake

trated in Fig. 5.11, and these do not take into consideration the presence of such features as marginal geosynclinal sediments, which could be extruded between the two continents like putty in the jaws of a vice. Island arc systems may also be involved which could complicate the structure of collision zones still further. The structure of each intra-continental cordillera must be examined individually therefore, and each of its component structural elements identified in order to untangle the geological history of the collision zone. The major problem with this process is identifying the individual elements.

The structure may be complicated by the involvement of 'flakes' of crust separated by the collision. Figs. 5.11B–C show flakes detached on one side and on both sides. Either situation will make the surface geology more complex. In the Eastern Alps, for instance, a flake from the southern plate has overridden the northern plate and trapped underneath itself considerable quantities of sediment. Whatever form the plate structure may take in these suture zones it is bound to be at least one and a half times the thickness of 'normal' continental plates.

Although sediments may be relatively superficial they are crucial in the identification of suture zones. Geosynclinal sediments may be trapped by the collision, but deformation in this zone may be so extreme as to obscure their original lithology. Compression may create such structures as folds, thrust faults and *nappes*. The rocks may have been subjected to either thermal or dynamic metamorphism, or both, so that even their texture and mineralogy may have been altered. New sediments such as *molasses* may have been formed at the time of orogenesis. (Molasse collectively describes a cumulate of fluvial sediments eroded from the rising mountain chain and deposited in fold traps as shown in Fig. 5.8.) The structural complexity may be such that it is often impossible to unravel the original stratigraphy of these sediments. It may, however, be possible to identify rock types and relate them to geological environments, for example, abyssal shelf sediments. The presence of high-pressure, low-temperature metamorphism may indicate the position of a former subduction zone. By tying together a great number of small scraps of geological data a composite picture may be built up which shows the evolution of each section of each mountain range (see Chapter 6).

Where two continents collide very little trace of original ocean floor may be found in the mountain chains. Cherts, pillow lavas and serpentinites are sometimes found together in small quantities in intra-continental cordilleras and this association of rock types is often known as the *Steinmann trinity*, after its discoverer. Three other lithologies—gabbros, amphibolites and unserpentinized ultrabasics—have also been recognized as part of this association. This entire association is sometimes referred to as the *ophiolite suite*, and is believed to constitute fragments of oceanic crust incorporated into the mountain chain by the collision process. These are the 'pods' illustrated in Fig. 5.8.

Because there are so many possible variations in continent/continent collision no 'typical' cross section or block diagram has been shown.

Island arcs in collision

In situations where island arcs collide with one another they may fuse together to form linear segments of continental crust. Regions which have been formed in this fashion may be identifiable by

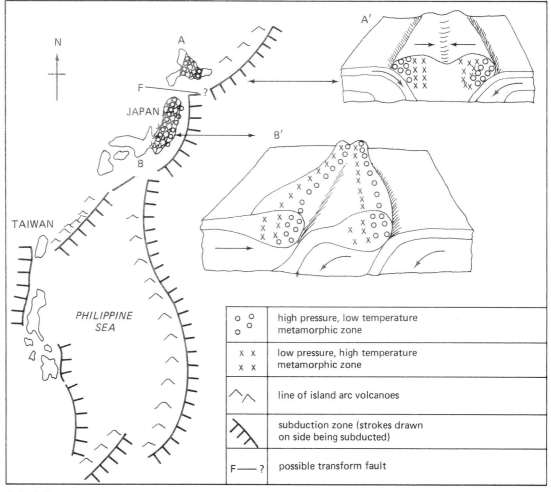

Fig. 5.12 Collision between island arcs in the area of Japan and the Philippine sea. A' and B' are sectional block diagrams showing the structure east/west through areas A and B

the presence of the paired metamorphic belts described on page 59 which are diagnostic of subduction zones. The plan of these zones may also give some indication of the structure of the collision which created them. The sketch map in Fig. 5.12 shows the arrangement of such zones in Japan. On the main island, Honshu, the sequence of zones from west to east is thermal, dynamic, thermal, dynamic. The inset sectioned block diagram B¹ shows the type of collision which could bring about this arrangement; two island arc systems colliding, where both subduction zones plunge in the same direction. The area on the map marked B may exemplify the structure shown in B¹. Honshu may have been formed by this type of collision. If the Philippine sea is closing due to the convergence of the island arc systems to the east and west this pattern may be extended southward in the future.

The arrangement of zones is different on the northern Japanese island, Hokkaido. There the sequence of metamorphic belts is dynamic, thermal, thermal, dynamic (A on map). This arrangement could have been produced by the situation in A¹ which shows two subduction zones plunging towards each other. It is suggested that a transform fault may run between Honshu and Hokkaido separating the two types of island arc collision. In Chapter 6 it will be shown how the earliest continental crust may have been built up by this type of collision.

6

Continent building

The structure and age of continents

Geological dating

While geology was developing as a science throughout the nineteenth century, the early pioneers had little knowledge regarding the actual ages of the rocks they were studying. They did, however, deduce the succession in which geological strata had been laid down and arrange them into a relative order (i.e. B succeeds A and precedes C), which enabled them to construct a stratigraphical column (Fig. 6.1). This column provided them with a *relative time scale*, which was subdivided into *eras* and *periods*, representing phases when sediments were deposited which had diagnostic characteristics of lithology and palaeontology. The subdivisions of the column were, therefore, erected on the basis of sedimentary succession and this is reflected strongly in the structure of the column. Younger sedimentary rocks are likely to be more accessible, less distorted, and to contain more complete fossil assemblages than older ones. They can therefore be studied in greater detail and subdivided more finely than their older counterparts. Accordingly, the more recent divisions of the stratigraphical column represent shorter periods of time than the older ones.

The most important division in the column was, not surprisingly, erected between those successions which were found to contain abundant fossils and those which were not. The term *Phanerozoic*, or 'time of life', was given to that part of the column lying above the base of the Cambrian. All successions below that were collectively labelled the *Pre-Cambrian*. It can be seen from Fig. 6.1 that most of the divisions of geological time occur in the Phanerozoic. Fig. 6.1 illustrates, however, how small a proportion of geological time is actually occupied by the Phanerozoic; 570 million years out of the 4600 million years since the formation of the Earth. Since most continental crust is capped with Phanerozoic deposits it is understandable that they should be the object of the bulk of geological research. It should be borne in mind, though, that the greater part of the mass of continental crust is made up of rocks which are Pre-Cambrian in origin, mostly unfossiliferous and usually extremely folded and faulted. These rocks cannot therefore be differentiated into periods in the same fashion as Phanerozoic rocks. In the absence of such a method of relative dating it is necessary to use some method of *absolute dating*.

In order to put absolute dates on the origins of particular rocks one must use methods for deducing, *quantitatively*, how old they are. Absolute age deductions are made by the study of certain *radioactive nuclides* found in some rock-forming minerals. Radioactive nuclides undergo a process known as *radioactive decay*. Uranium 238, for instance, decays to form lead 206. In this instance U238 is called the parent nuclide and Pb206 is called the *daughter nuclide*. The time taken for half the original number of parent nuclides to transform to daughter nuclides is called the *half-life* of that particular decay. Each particular radioactive decay

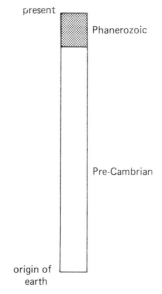

Fig. 6.1 The Phanerozoic as a proportion of geological time

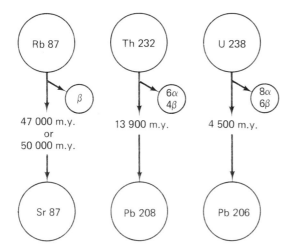
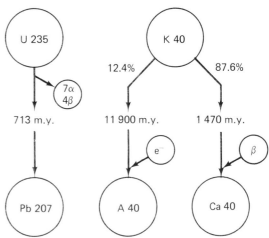

Fig. 6.2 Radioactive decays used for geological dating

has its own specific half-life. The number of parent nuclides transforming to daughters per unit time is called the *decay constant* (λ) of a reaction.

Where the half-life, or decay constant, is known for a particular decay and the ratio of parent/daughter nuclides present in a sample is found by analysis with a mass spectrometer, the absolute age of a sample can be determined. This presupposes that certain conditions are met. These are (i) that allowances can be made for any additional daughter nuclides originally incorporated into the sample; (ii) that the sample becomes a closed system fairly rapidly after it has been formed and (iii) that this system remains closed up to the time of analysis. (By the term 'closed system' it is meant the sample is in such a condition that it does not lose or gain parent or daughter nuclides except from the actual decay reaction.)

Minerals which meet these conditions and whose origin may thus be dateable may include those formed in igneous rocks which have cooled from a magma; those formed during a metamorphic event; and those formed during, or shortly after, the deposition of sediments. In these cases the dates produced will be late in cooling, late in metamorphism and late in diagenesis respectively, marking the dates when the minerals became effectively closed systems. These systems may be re-opened by later metamorphic events, so that the radio-chronological age of a sample may be that of its last metamorphism and not its original date of formation.

Five decay reactions which are commonly used in geochronology are represented in Fig. 6.2. These are the decays of uranium 238 to form lead 206; uranium 235 to form lead 207; thorium 232 to form lead 208; rubidium 87 to form strontium 87; potassium 40 to form argon 40 and calcium 40. All of these are nuclides found in minerals and having half lives which are long enough to allow them to be of use in geochronology.

Age/structural provinces on continents

The hearts of continents are composed of great slabs of hard, crystalline, igneous and metamorphic rocks. These slabs are Pre-Cambrian in origin and are known as shields or *cratons*. They are also sometimes called *platforms*, or *basements*, since they form the foundations underlying much of the younger, Phanerozoic deposits. Those areas of the world where Pre-Cambrian rocks are exposed at the surface are delineated on the map in Fig. 6.3. Large tracts of basement are overlain by Phanerozoic deposits which are relatively flat-lying. Other Phanerozoic deposits have been compressed into cordilleras marginal to shields, or trapped between adjacent shields. (See also Fig. 2.5.)

By dating samples taken from Pre-Cambrian basements and by analysing their structural framework it has been possible to subdivide shields into geological zones. These zones, defined on the basis of age and structure, are known as *provinces*. The provinces of the Canadian shield are shown in Fig. 6.3 (inset). (Provinces of similar age have been similarly shaded.) Provinces younger than 2000 million years tend to be linear in structure and occupy well-defined zones. They may represent the eroded roots of ancient orogenic belts. The older provinces, often called the Archean shields, show no such structure, and their origin is more problematic.

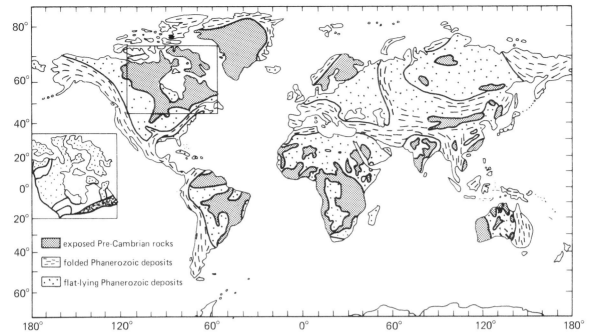

Fig. 6.3 World distribution of exposed Pre-Cambrian rocks. (Inset: structural provinces of the Canadian shield.)

The Growth of Continents

The origin of continental crust

Whilst the origin of oceanic crust can be observed at mid-ocean ridges and is fairly well understood, the origin of continental crust is more obscure and thus more controversial. The structure of continental crust is complex and its history goes back more than 3800 million years (the radiometric date on the oldest known rocks, found in Archean terrain in Greenland). There are two schools of thought on the origin of sial: (i) that it was chemically differentiated early in Earth's history and has since been continually reworked by tectonic processes; (ii) that it is continually being created at convergent plate margins and therefore that continents have continually grown throughout geological time. Recent evidence based on analyses of the Sr 87/Sr 86 ratio in crustal rocks indicates that the latter school of thought is more likely to be correct.

The process suggested for creating new continental crust is illustrated in Fig. 6.4. As a subducted plate descends into the mantle it starts to remelt. The first minerals to melt will be those with the lowest melting points. Melting points will be lowered still further by the presence of water from subducted wet sediments. Early melting minerals are usually those which also have the lowest densities. As the slab partially melts the less dense molten minerals will percolate up through the denser unmelted material and separate out as *diapirs* of magma. Meanwhile the residue, its average density now increased, will continue to plunge downwards. Being depleted of light, silica-rich minerals the residue is ultrabasic. This process is known as *fractional melting differentiation*.

The diapirs of magma, having low densities, continue to ascend and emplace themselves in the overriding plate. When they cool they become an integral part of the continent and thus add to its bulk. By this means continental crust is created above subduction zones.

Archean shields

The rocks of Archean shields are characterized by geological provinces occurring in whorled patterns rather than linear zones. They do, however, have a discernible structure: this is one of discontinuous strips of greenstone enclosed by vast terrains of gneiss and intruded by granites and granodiorites (Fig. 6.5). The gneisses, granites and granodiorites of Archean shields bear a strong resemblance to the batholithic intrusions which overlie continental marginal subduction zones (page 61). The greenstone belts are composed of metamorphosed sedimentary and volcanic deposits. The chemistry of the volcanics is intermediate between that of

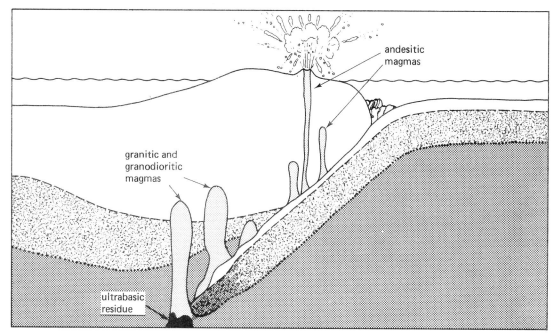

Fig. 6.4 Fractional melting differentiation

island-arc andesites and mid-oceanic tholeiites. For this reason it has been suggested that greenstone belts represent the remnants of retro-arc basins (page 59). Tholeiitic magmas are emplaced in fissures in basin floors and andesitic lavas are erupted from island-arc volcanoes.

One suggested mode of origin for Archean greenstone/gneiss terrains is illustrated in Fig. 6.6. A marginal basin develops and sediments and lavas accumulate on its floor (a); a compressive phase ensues and the basin deposits are folded with intrusions of granite/granodiorite (b); a second arc and basin system develops (c) and is subsequently compressed in the same manner as the first (d); a third basin develops and the cycle re-starts (e). The structure thus produced bears a strong resemblance to the Archean terrain illustrated in Fig. 6.5. This may, therefore, be the process which produced the Earth's oldest areas of continental crust.

Accretion of orogenic provinces

Provinces which are younger than 2000 million years tend to be elongate and to fringe older shield areas. This structure may be explained if these provinces are seen as the eroded roots of ancient cordilleras, formed up against the older shield margins. They may thus be termed *orogenic provinces*.

Continental growth may be seen as a process of accretion, the Archean shields being formed by the fusion of marginal basins and cordilleras being added to the margins of these shields during subse-

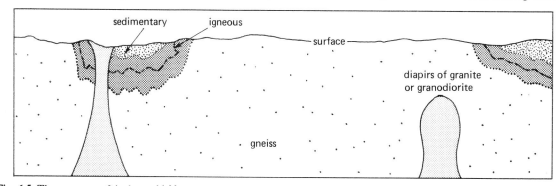

Fig. 6.5 The structure of Archean shields

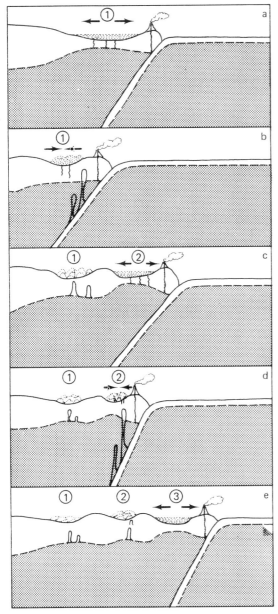

Fig. 6.6 Formation of Archean shields by the collision of island arcs and retro-arc basins

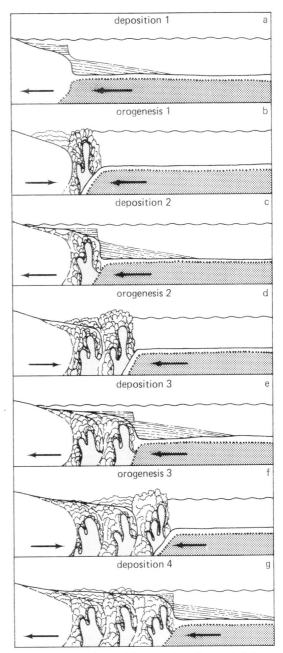

Fig. 6.7 Continental growth by accretion of orogenic belts

quent periods of compression. The repetition of this process causes the episodic growth of continents, as represented in Fig. 6.7. Periods of deposition alternate with periods of orogenesis, respectively representing periods when the continental margin is under tensional or compressional stress. A succession of geosynclines thus become incorporated in the plate margin. When eroded down the roots of each of these will form a new orogenic province, extending the continental craton. This process is a combination of reworking of continental crust (the 'geosynclinal' sediments originally came from the continents) and generation of new continental crust (by fractional melting differentiation of the subducted plate). The structure of even the oldest continental crust may thus be explained in terms of plate tectonics.

Most continental cratons are constructed from

central Archean shield terrains, flanked by progressively younger orogenic provinces and partially overlain by younger deposits. Younger orogenic belts which appear to cut through the centres of shields in fact usually prove to be trapped between adjacent shields. Larger continental cratons can therefore be created by collisions between smaller ones. The cordilleras marking the suture zones usually betray their origins by the presence of fragments of the ocean floors which had been all but destroyed by the collisions.

Continental movement and growth

The frequency distribution of age determinations on Pre-Cambrian basement rocks (Fig. 6.8) indicates that continent building is an episodic process. The peaks on the graphs indicate phases of formation. The earliest known phase of continent building took place between 3800 m.y. and 3500 m.y. ago, but this probably creates less than 10 per cent of present continental crust. As much as 60 per cent was formed during the second great phase 2900 m.y. to 2600 m.y. ago. Continental crust created up to this date appears to be of the Archean type described above. In subsequent phases 1900 m.y. to 1700 m.y. ago and again 1100 m.y. to 900 m.y. ago the crust created was of the orogenic type, in linear provinces.

The most recent phase of continent building started about 600 m.y. ago and so is not shown on the Pre-Cambrian graph. A more detailed picture of continental movement and growth has been compiled for this period. This has been done by retracing the movements of continents throughout the Phanerozoic and tying in mountain building with continental advances and collisions. The co-positioning of continents at various points in geological history has been done partly by determining the positions of the north and south poles in magnetized rock samples of the appropriate ages and resiting them to fit in with the Earth's present magnetic field. This is possible for the Phanerozoic since all continents are covered to some extent with relatively undisturbed deposits younger than Pre-Cambrian. More recent (less than 200 m.y.) conti-

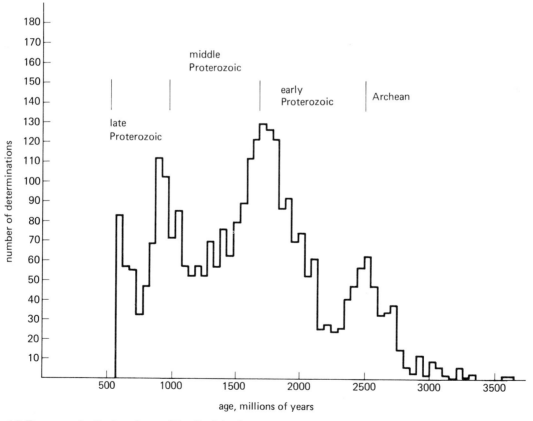

Fig. 6.8 Frequency distribution of ages of Pre-Cambrian basements worldwide

nental movements can be retraced much more accurately by studying the magnetic anomaly patterns of the ocean floors. The matching of the features mentioned on pages 15–24 is also important in the work of re-positioning continents.

One suggested scheme of continental movement from the Cambrian through to the present is illustrated in Fig. 6.9. During the Cambrian there existed five continental cratons, for convenience called the North American, European, Siberian, and Chinese continents, and a huge southern continent, *Gondwanaland*. By the Ordovician period Europe and America had collided to form *Euramerica*, the suture being marked by the *Caledonide* mountain chain. The Sayan, Yablonovy and Stanovay ranges were created when the Siberian and Chinese continents collided to form the Asian continent during the Silurian/Devonian. Asia and Europe started to collide about this time to form the *Uralide* chain and this orogeny continued up to the Permian. The Northern supercontinent thus created is known as Laurasia. During the Carboniferous period, Europe and the United States area were in the tropical zone and covered by large tracts of coal-forming swamp. At the same time Gondwanaland was straddling the South Pole and undergoing extensive glaciation. Towards the end of the Carboniferous period, Gondwanaland, which had been moving steadily northward, collided with Laurasia creating the *Hercynide* mountain chain. The effect of this collision was that all the world's continents were welded together to form a supercontinent, known as Pangaea.

The breakup of Pangaea commenced some 200 m.y. ago during the Triassic. By the end of that period it had separated once more into Laurasia and Gondwanaland. The latter had itself started to fragment into the southern continents as we know them today. South America and Africa were virtually separated by the end of the Cretaceous, with Africa (including Italy and Arabia) rotating anticlockwise on a collision course for Europe. This collision, lasting through the Tertiary, threw up a mountain chain stretching across southeastern Europe known as the *Alpine Orogeny*. This mountain-building activity was stretched still further eastwards when India collided with southeast Asia to create the Himalayas.

These palaeogeographic maps essentially show the continental cratons as they are today, relocated at various points in geological time by means of palaeomagnetic and other data. It should be borne in mind, however, that, since continents are actually built by these movements, there is an inherent flaw in this exercise, which is that some areas shown on earlier maps had not been created by the periods illustrated. However, the maps do act as convenient frames of reference for some geological studies. A more accurate picture of how continent building operates can be acquired by means of a local study.

A local study: the building of Britain

The oldest known rocks in Britain are those which originally comprised parts of the ancient European and North American continents. These rocks are today found as a sliver of the northwest corner of Scotland and as a basement underlying the younger rocks of Southeast England. When these two continents collided together during the middle Palaeozoic they created a cordillera which built parts of Scandinavia, Greenland, Newfoundland and Canada, and a considerable tract of the British Isles (Fig. 6.10). This event was the *Caledonian orogeny* and its line is reflected in the strong northeast/southwest structural trend of the Lower Palaeozoic rocks of the British Isles. Throughout the Devonian the forces of erosion attacked the Caledonian Mountains and the fluviatile and deltaic deposits produced formed the molasse we today call the *Old Red Sandstone*. At that time Britain lay some 20°–30° south of the equator and the redness of the sandstone indicates an arid climate. (The area would also have lain in the rain shadow of the Caledonian Mountains.)

The southern edge of the Euramerican continent formed a geosyncline where marine deposits were laid down throughout the Devonian and Lower Carboniferous. The Caledonian mountains had been worn down and shallow shelf seas transgressed the southern margin of Euramerica by the Lower Carboniferous period. The area had by that time moved into the tropical rain belt and coal-forming swamps abounded. Many fluctuations in sea level caused tracts of land to alternate between being areas of swamp, and being areas of shallow sea. This produced the *cyclothem* rock sequence seen in Carboniferous rock strata.

Gondwanaland moved up from the south to collide with Euramerica during the Carboniferous. The cordillera thus produced, created the *Appalachian* belt of America, parts of North Africa and Central Europe and the southwest corners of Great Britain and Ireland. This was the *Hercynian orogeny* and it imparted a dominant east-west structural trend to the areas it created (Fig. 6.11). The Hercynian mountains provided the source material

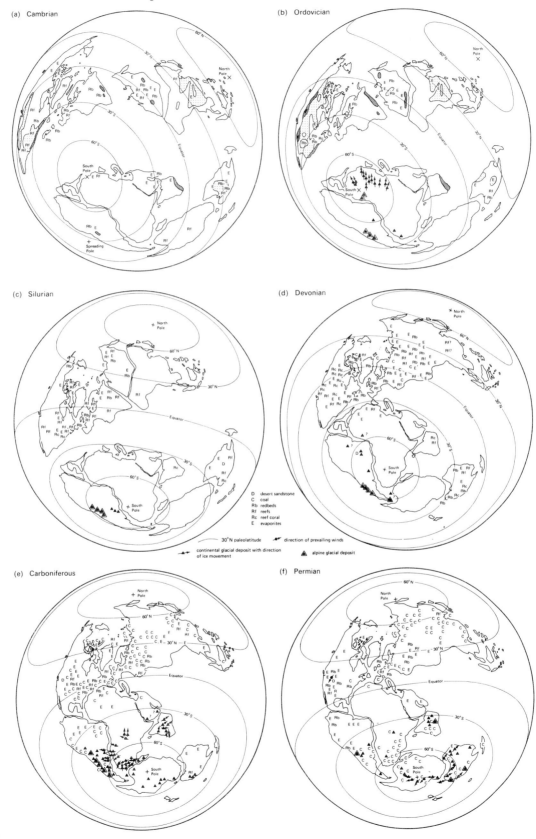

Fig. 6.9 Continental movement throughout the Phanerozoic
(a) Cambrian (b) Ordovician (c) Silurian (d) Devonian (e) Carboniferous (f) Permian

D desert sandstone
C coal
Rb redbeds
Rf reefs
Rc reef coral
E evaporites

— 30°N paleolatitude
— direction of prevailing winds
▲▲ continental glacial deposit with direction of ice movement
▲ alpine glacial deposit

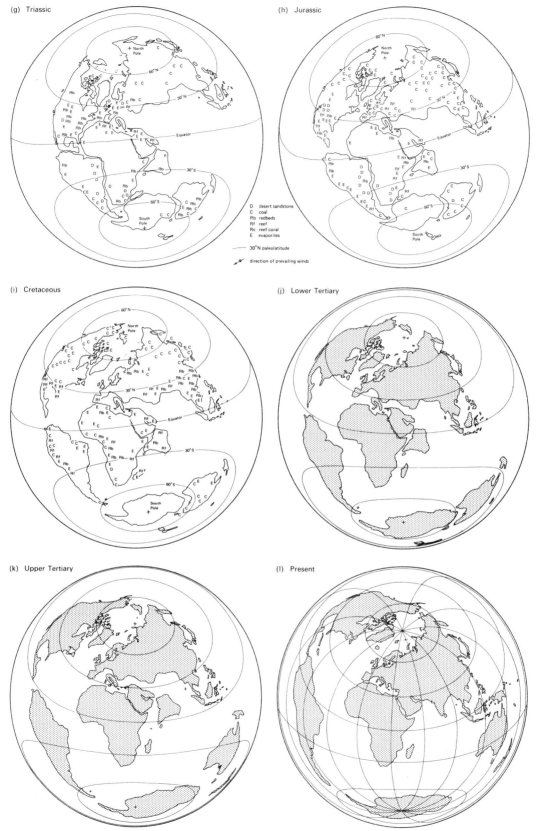

75

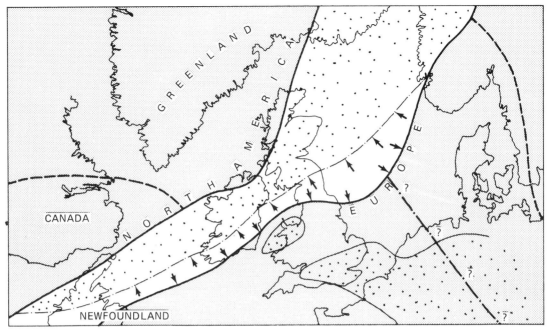

Fig. 6.10 The building of Britain. The Caledonian orogeny: continental cratons are shaded grey; ocean crust is unshaded; areas of deposition are stippled; broken lines separate provinces; arrows mark subduction zones; question marks indicate areas of doubt

for the *New Red Sandstone* of the Permian, which was created in a similar environment to the Old Red Sandstone. The reduction of the land level by erosion continued through to the Triassic, to produce a low-lying arid plain, upon which were deposited evaporites and marls. By the Jurassic period marine transgressions had turned parts of the area into shelf seas where shales and limestones were deposited. The area was then 30° north of the equator and corals abounded in the warm shallow water. During the Jurassic the opening of the Atlantic Ocean tore apart Europe from North America, dividing the Caledonian and Hercynian orogens.

This established the basic structure of Britain as it is today, since it was not involved directly in any more continent building. Throughout the Cretaceous period sea levels rose and *chalk* was deposited in Southeast England. During the Triassic period Africa collided with southern Europe creating the *Alpine orogen* and although Britain was not directly involved, the impact gently folded the rocks of Southeast England, producing the structures of the Downs and the Weald.

Fig. 6.11 Caledonian and Hercynian structural belts on a pre-Alpine refit map

Recommended Further Reading

Earth History and Plate Tectonics by Carl K. Seyfert and Leslie A. Sirkin, (Harper & Row, New York, 1973)

Understanding the Earth edited by Gass et al, (Artemis Press, Sussex, 1972)

The Plain Man's Guide to Plate Tectonics by E. R. Oxburgh

The Geological Evolution of Europe by D. V. Ager Reprints from the Proceedings of the Geological Society of London: obtainable from Publications Secretary, Geologists Assoc, Dept of Geology, Imperial College, London SW7 2AZ

British Stratigraphy by F. A. Middlemass (Allen & Unwin, 1974)

Principles of Physical Geology by A. Holmes (Nelson, 1965)

Readings in the Earth Sciences (2 vols), Scientific American Offprints 801–874 (W. H. Freeman, San Fransisco)

Index

absolute dating, 67
Abyssinian Rift, 43, 44, 45
accretion, 63
Aden Rift, 43, 45
Afar triangle, 46, 47, 53
Ahmed, F., 15, 17
alkali basalt lavas, 44, 45, 54
alkali metals, 44, 45
Alpine orogen, 76
Alpine orogeny, 73
Alps, 58, 65
amphibolites, 65
Andes, 31, 58, 61, 64
andesite, 58, 62, 63, 69
Antrim Plateau, 47
Appalachian belt, 73
Archean shields, 68, 69, 70
argon, 68
arid regions, 21
asthenosphere, 34, 35, 36, 37
Atlantic Ocean, 26, 29, 45, 46, 48, 57
atmosphere, 5
atolls, 53, 54
axis of rotation, 37, 38

Baja California, 63
Ballantrae, 16
basalt, 28, 39, 45
basements, 15, 68
batholiths, 61, 62, 69
beds, 52
Benioff, Hugo, 31, 32
Benioff Zones, 31, 32, 44, 58, 59, 63
Black Forest horst, 43
blueschist metamorphism, 58
Bullard, Sir Edward, 14, 15

calcium carbonate, 19
Caledonian orogeny, 73
Cambrian, 67
Canadian shield, 68, 69
Carboniferous, 20, 21, 73, 74
Carlsberg Ridge, 28
chalk, 76
chert, 56, 65
Cloos, Hans, 42, 44
coal, 21, 22, 73

compression, 34, 35
conglomerate, 52
continental crust, 11, 26, 35, 57, 69
continental plates, 35
continental shelves, 49, 50, 51
continental slopes, 49, 52
convection cells, 32, 37, 38
convergence of species, 21, 22, 23
convergent boundaries, 37, 57
coral, 19, 53, 74, 75, 76
cordillera, 31, 32, 37, 60, 64, 70, 76
core, 7, 10
cratons, 68, 71, 72, 73
Cretaceous, 75, 76
crust, 7, 11, 12, 35
crustal block mountains, 43
crustal extension, 59
crustal separation, 44
Curie temperature, 17
cyclothem, 73

Dana, James, 54
Darwin, Charles, 54
daughter nuclide, 67, 68
decay constant, 68
Deccan Plateau, 48
de-coupling, 35
deep-ocean floor, 50
density layering, 5
Devonian, 73, 74
diapirs, 69, 70
divergence of species, 21, 22, 23
divergent boundaries, 36
driving mechanisms, 37, 39
ductile necking, 51
dune-bedded sandstones, 20
dynamic metamorphism, 58, 66

earthquakes, 7, 27, 28, 30, 31, 32, 52
East African Rift Valley, 44, 45
East Pacific Rise, 63
ensimatic sediments, 61
epicentre, 8, 27, 30, 31, 32
eras, 67
eugeosyncline, 61, 62
Euramerica, 73
evaporites, 74, 75, 76

Everest, Mount, 31

failed arms, 45
faulting, 35
flakes, 65
flysch, 51, 52
focus, 8, 31, 32
folding, 35
fold mountains, 51
fossil magnetism, 28
fossils, 21, 52, 67
fractional melting differentiation, 69, 71

geochronology, 68
geosynclinal couplet, 61
geosyncline, 51, 52, 60, 61, 71
geothermal heat, 26, 32, 36, 53
geothermal turbulence, 37, 39
Giant's Causeway, 48
glaciation, 13, 19, 21, 22
Glomar Challenger, 29
gneiss, 69
Gondwanaland, 73
graben, 43 52, 53
graded beds, 52
granite, 61, 69
granodiorite, 61, 69
Greenland, 69
greenschist metamorphism, 59
greywackes, 61
guyots, 53
gypsum, 21, 22

half-life, 67, 68
Hawaii, 7, 54
Hercynian orogeny, 75
Hercynide chain, 74
Hess, Professor H., 26, 27, 31, 32, 35
Hesse Rift, 43
Himalayas, 64, 73
Hokkaido, 66
Honshu, 65, 66
horsts, 43
'hot spots', 42, 44, 54
hydrosphere, 6

Iceland, 28, 29, 52
igneous rocks, 7, 63

Japan, 65, 66
Jurassic, 73, 75

Kamchatka-Kurile arc, 31
Klamath Mountains, 63

Koppen, 21, 22
Kuril Trench, 32

land bridge, 22, 23, 24
laterites, 21
Laurasia, 73
lava, 6, 44, 45, 53
lead, 67
limestones, 52
lithosphere, 6, 34
low velocity zone, 12, 34

MacDonald Seamount, 54
magma, 7, 26, 36, 37, 39, 69
magnetic banding, 28, 29
magnetic surveys, 28
manganese nodules, 56
mantle, 7, 10, 32, 34
Marginal basins, 59, 61
Mariana Trench, 31
Matthews, D. H., 28
membrane tectonics, 39
Mesosaurus, 18, 21, 22
metamorphism, 58, 61, 62, 65, 68
meteorites, 6, 7
mid-Atlantic-Ridge, 26, 34, 52
miogeosyncline, 61, 62
Mohorovicic discontinuity (Moho), 11, 34
molasse, 62, 65, 73

New Red Sandstone, 76
normal faulting, 35, 36

oceanic crust, 11, 26, 34, 57
oceanic plates, 35
Old Red Sandstone, 73
ooze, calcareous, 56
ophiolites, 62, 65
ophiolite suite, 65
Ordovician, 74
orogenesis, 58
orogenic belts, 15, 64, 68, 72, 73, 76
orogenic provinces, 70

P waves, 8, 9
Pacific Ocean, 30, 31, 54, 55, 57, 63
palaeoclimates, 18
palaeomagnetism, 16–17
palaeontology, 21
Pangaea, 15, 16, 64, 73
parent nuclide, 67, 68
pelagic sediments, 51, 56, 63
peridotite, 7
periods (geological), 67

Permian, 16, 20, 21, 73, 74
Peru/Chile Trench, 31, 61
Phanerozoic, 67, 69
Phillipine Sea, 66
pillow lavas, 53, 65
Pitcairn Island, 54
plankton, 56
plate, crustal, 34, 35
plate boundaries, 35
plateau basalts, 45, 47, 48
platforms, 68
plumes, 39, 42, 44, 54
plutons, 61
pods, 62, 65
polar reversals, 28, 29
polar wandering, 17
polar wandering curve, 17, 19
potassium, 68
Pre Cambrian, 67, 68, 69
pressure waves, 8, 9
primary waves, 8, 9
Proterozoic, 72
provinces, 68
pyrolite, 10

radioactive decay, 67
radioactive nuclides, 67, 68
radiolaria, 56
red clays, 56
Red Sea Rift, 43, 45, 46, 53
reefs, 51, 54, 74, 75
reflected waves, 8, 9
refracted waves, 8, 9
retro-arc basins, 59, 61, 69
reversals (polar), 11
reverse faulting, 35, 36
Reykjanes Ridge, 28, 29
rheosphere, 12
Rhine Rift Valley, 43
ridge, mid-oceanic, 26, 28, 29, 32, 34, 36, 52
ridge segments, 37
rift valleys, 42, 43, 46, 52
rifting, 36, 42, 43
rock magnetism, 17
rock salt, 21, 22
rubidium, 68

San Andreas Fault, 63
sedimentation, 29, 51, 52, 59, 60, 61, 62, 65, 73, 76
sediments, 27, 49, 52, 61, 62, 65, 73, 76
seismic event, 8
seismic patterns, 35
seismic profile, 34
seismic surveys, 26

seismograph, 8
serpentinites, 65
shadow zone, 9
shale, 52
shields, 15, 68
sial, 11
Silurian, 73
sima, 11
Snider, Antonio, 13
spreading rates, 37, 59
Steinmann trinity, 65
stratigraphy, 16, 18, 67
stress, 35
strike-slip faulting, 35, 36, 63
strontium, 68
subduction, 32, 36, 37, 58, 60, 65, 69
Sunda, 64
swell and basin topography, 44, 45
sylvite, 21

Taylor, F. B., 13
Tazieff, Haroun, 53
tension, 35, 36
Tertiary, 73, 75
thermal metamorphism, 59, 66
thermistor probe, 26
tholeiitic basalts, 45, 47, 48, 52, 58, 70
thorium, 68
thrusting, 35, 36
tillites, 18, 19, 21
time scale, relative, 67
torsion, 35, 36
transform faults, 26, 36
trenches, deep ocean, 30, 31, 32, 34
Triassic, 20, 75, 76
Tristan da Cunha, 54
turbidity currents, 52

ultrabasics, 65, 69
upwarping, 43, 44
Urals, 64
uranium, 67, 68

Vine, F. J., 28
viscous drag, 42
volcanic island arcs, 31, 58
volcanic islands, 29, 53, 54, 55
volcanoes, 6, 29, 30, 31, 32, 69
Vosges horst, 43

Walvis Ridge, 54
Weald, 76
Wegener, Alfred, 13, 21, 22
wrenching, 35, 36, 63

xenoliths, 7